Bacillus thuringiensis:
A Cornerstone
of Modern Agriculture

Bacillus thuringiensis: A Cornerstone of Modern Agriculture has been co-published simultaneously as *Journal of New Seeds*, Volume 5, Number 1 and Numbers 2/3 2003.

Bacillus thuringiensis: A Cornerstone of Modern Agriculture

Matthew Metz, PhD
Editor

Bacillus thuringiensis: A Cornerstone of Modern Agriculture has been co-published simultaneously as *Journal of New Seeds*, Volume 5, Number 1 and Numbers 2/3 2003.

Routledge
Taylor & Francis Group

NEW YORK AND LONDON

Bacillus thuringiensis: A Cornerstone of Modern Agriculture has been co-published simultaneously as *Journal of New Seeds*, Volume 5, Number 1 and Numbers 2/3 2003.

First published 2003 by The Haworth Press, Inc.

Published 2021 by Routledge
605 Third Avenue, New York, NY 10017
2 Park Square, Milton Park, Abingdon, Oxon OX14 4RN

*Routledge is an imprint of the Taylor & Francis Group,
an informa business*

Library of Congress Cataloging-in-Publication Data

Bacillus thuringiensis : a cornerstone of modern agriculture / Matthew Metz, editor.
   p. cm.
  "Co-published simultaneously as Journal of new seeds, volume 5, number 1 and numbers 2/3 2003."
  Includes bibliographical references and index.
  ISBN 1-56022-108-9 (Hard Cover : alk. paper) – ISBN 1-56022-109-7 (Soft Cover : alk. paper)
  1. Transgenic plants. 2. Agricultural pests–Biological control. 3. Bacillus thuringiensis–Toxicology. I. Metz, Matthew. II. Journal of new seeds.
SB123.57.B33 2003
632'.96–dc21

                  2003011498

ISBN 13: 978-1-56022-108-1 (hbk)

Bacillus thuringiensis: A Cornerstone of Modern Agriculture

CONTENTS

ABOUT THE EDITOR

Matthew Metz, PhD, studied Biology at the California Institute of Technology, graduating with a BS in 1995. He earned his PhD in the Plant Biology department at the University of California, Berkeley in 2001. His graduate research, on molecular genetics of plant disease resistance and on *Xanthomonas* and *Pseudomonas* bacterial pathogens of plants, was supported by a Regents Fellowship. Dr. Metz did research on *Agrobacterim*-mediated transformation of plants in a postdoctoral position at the University of Washington, Seattle in 2001 and 2002. He is currently at the U.S. Agency for International Development (USAID), in the Bureau for Economic Growth, Agriculture and Trade. His work at USAID is through an American Association for the Advancement of Science fellowship. He is currently helping to administer programs at USAID that fund scientific research in developing countries and in co-operative projects between scientists of Middle Eastern countries. His research has continually involved genetic engineering and agriculture. He has also taken on numerous public policy and scholarly activities that address these topics, including testimony before government hearings, and service on an American Academy of Microbiology colloquium, including co-authoring the colloquium report, entitled, '100 years of *Bacillus thuringiensis*: a critical scientific assessment.'

Foreword

Huge damage has been inflicted on crops by pestiferous insects throughout the world since man first became a farmer. Due to the emergence and spread of insecticide resistance, concern with environmental pollution, and the high cost of new chemical insecticides, beneficial natural pathogens have attracted increasing attention. There is an urgent need for a biological agent that possesses desirable properties of a chemical pesticide such as suitability for industrial scale production, long shelf life, convenient application using conventional equipment and transportability, but with high specific toxicity to target organisms and safety to non-target organisms. The entomocidal bacterium, *Bacillus thuringiensis* (Bt) has provided a suitable candidate and the subject of intense investigation.

A century after the discovery of Bt, these spore-forming bacteria have become the most widely used biopesticides in biological control of insect pests of agriculture, forests and public health. During the past 20 years the exponential growth in our knowledge about the genetics and molecular biology of the organism, and advances in gene manipulation have enabled expression of insecticidal crystal proteins (ICPs) in transgenic crops, resulting in pest-resistant plants.

. This book is an extensive, unbiased compendium of current topics involving Bt as conventional biopesticides and in transgenic crops. Important information on key issues surrounding the controversial subject of genetically modified plants is dealt with throughout the volume. The authors possess a vast amount of knowledge and experience in the relevant topics and have meticulously gathered and integrated the experience and knowledge of others to compose a unique, accurate and

[Haworth co-indexing entry note]: "Foreword." Margalith, Yoel. Co-published simultaneously in *Journal of New Seeds* (Food Products Press, an imprint of The Haworth Press, Inc.) Vol. 5, No. 1, 2003, pp. xiii-xv; and: *Bacillus thuringiensis: A Cornerstone of Modern Agriculture* (ed: Matthew Metz) Food Products Press, an imprint of The Haworth Press, Inc., 2003, pp. xi-xiii. Single or multiple copies of this article are available for a fee from The Haworth Document Delivery Service [1-800-HAWORTH, 9:00 a.m. - 5:00 p.m. (EST). E-mail address: docdelivery@haworthpress.com].

comprehensive treatise on Bt: associated risks vs. benefits, development strategies and potential in world food production.

The volume consists of 12 articles organized into risks, benefits, resistance management and crop-specific case studies, and delves into the whole gamut of topics relating to Bt biopesticides and transgenic crops, including molecular biology techniques. The first article is an excellent account of extensive safety testing and a history of human exposure without adverse effects, proving Bt to be an important reduced risk pesticide. The second article, written by a researcher well versed in the fundamental science and development of microbial insecticides, provides evidence for safety toward non-target organisms based on knowledge of Cry protein specificity. It concludes that Bt crops can serve as the cornerstone for more environmentally sound integrated pest management programs.

Both the third and fourth articles provide comprehensive accounts of strategies to develop transgenic crops and to manage resistance, including the concept of gene pyramiding where such crops contain multiple ICPs in a single hybrid. The fifth article discusses research conducted in the Philippines, where a large number of transgenic Bt-rice varieties have been produced using the established transformation methods, *Agrobacterium*-mediated, biolistic and protoplast. Solutions to pest problems in worldwide potato production, with special reference to the potential of Bt-related biopesticides and transgenic crops, are discussed in the sixth article. The seventh and eighth articles address, respectively, environmental impacts of transgenic crops and requirements for resistance management using Bt cotton as a case study. The ninth article discusses a new method in a similar system, developed to assay gene expression using *Agrobacterium*; it will be extremely useful in evaluating gene constructs for input traits.

The tenth article describes a transgenic cauliflower expressing *cry1C* to control its cosmopolitan pests, diamondback moth and cabbage loopers, which have developed high levels of resistance to many chemical pesticides, including Bt biopesticides (in diamondback moth). Sugarcane producing Cry1Ab was developed to control the sugarcane borer, *Diatrea saccharalis,* by co-bombardment transformation (article 11) conferring resistance to the pest while conserving the desirable crop yield and quality. Finally, a case study of the benefits actually derived from Bt transgenic cotton in Argentina is described in the last article.

This book is an excellent source of information on the topics relevant to Bt development, utilization, resistance management and potential. It elaborates on the important issues facing scientists, legislators and consumers in balancing risks and benefits of Bt biopesticides and Bt crops. Safety of Bt conventional biopesticides to humans and non-target organisms has been proven over decades of use, providing reassurance that transgenic crops expressing Bt target specific proteins will also prove safe to human health and to the environment. The benefits we stand to gain from developing and introducing Bt crops include much needed increased food production, reduced agricultural input costs, decreased use of chemical poisons with subsequent diminished environmental pollution, and wide availability to developed and developing countries alike.

The articles in this volume provide a comprehensive and accurate account of the current status of risks and benefits of Bt-based pesticides and transgenic crops including rice, potato and cotton. Its primary beneficiaries will be students, scientists and professionals dealing with pest control, environmental protection and regulation of transgenic crops. Society as a whole stands to gain from increased understanding of the risks versus benefits of Bt transgenic crops and their inevitable introduction into integrated pest management programs for reducing conventional pesticide use and increasing production of food and fiber in all parts of the world.

Yoel Margalith
2003 Tyler Prize Laureate
Professor of Entomology
Director, Center for Biological Control (CBC)
Ben Gurion University of the Negev, Israel

Preface

Selective breeding has generated an amazing array of plant varieties suited to production of food and fiber for humans. Over millennia, crop varieties have been a target for exploitation by a host of opportunists that have also found the fruits of this labor to be an attractive food source. These 'pest' organisms have been an antagonistic force in the struggle to raise crop productivity. Pest control has been among the many innovations that have sustained continual increases in crop productivity including irrigation, fertilizers and advanced breeding programs.

While the term pesticide invokes for many the concept of a broadly toxic, synthetic chemical, the Gram-positive bacterium *Bacillus thuringiensis* (Bt) serves as a striking counterexample. This work seeks to illustrate the impact of Bt on pest control technology, and some of the related issues that surround its use. Development, use, and management of Bt derived technologies are covered in this set of papers, through a series of reviews and original research articles. Deployment of genetically engineered crops has been challenged by many health, economic, environmental, and social concerns. Many of these concerns, but not all, can be addressed with the material presented here.

Since its discovery about a century ago, Bt has developed into an important tool for pest control. Unlike many synthetic chemicals, Bt's mechanism of insecticidal activity is highly specific. Certain strains of the bacterium, and the associated insecticidal crystal proteins (ICPs), are effective against only certain insects. Typically the range of activity for any single Bt isolate is restricted to one of the orders Lepidoptera

The views expressed in this article are those of the author and do not necessarily indicate a particular position of the US Agency for International Development.

[Haworth co-indexing entry note]: "Preface." Metz, Matthew. Co-published simultaneously in *Journal of New Seeds* (Food Products Press, an imprint of The Haworth Press, Inc.) Vol. 5, No. 1, 2003, pp. xvii-xxiv; and: *Bacillus thuringiensis: A Cornerstone of Modern Agriculture* (ed: Matthew Metz) Food Products Press, an imprint of The Haworth Press, Inc., 2003, pp. xv-xxii. Single or multiple copies of this article are available for a fee from The Haworth Document Delivery Service [1-800-HAWORTH, 9:00 a.m. - 5:00 p.m. (EST). E-mail address: docdelivery@haworthpress.com].

(butterflies/moths), Coleoptera (beetles), or Diptera (flies/mosquitoes). This specificity underlies one of the huge advantages of Bt: its safety to humans. As reviewed in the article by Kough, a history of regulated use has proven Bt to be one of the safest products for pest control in terms of impact on human health.

The specificity of Bt renders it not just safe for humans, but also to a plethora of what are termed 'non-target' organisms. The broad toxicity of many pest control treatments has serious detrimental consequences for organisms exposed to the residues in and emanating from treated fields. Typically, organisms can absorb toxic chemicals in a variety of different ways. However, Bt is effective only upon ingestion, and then is toxic only to vulnerable organisms over a narrow range of species. As covered in detail by Federici (pp. 11-30), harm from Bt is generally restricted to targeted pests because of both target organism specificity and the fact that it must be eaten. The safety of Bt to the vast majority of non-target organisms contributes greatly to reducing environmental damage posed by the intrinsically disruptive practice of agriculture. Furthermore, Bt leaves predators and parisitoids of agricultural pests in a viable state to provide natural pest protection and integrated pest management (IPM) tools that are often destroyed by chemical insecticides.

Bt is not only biologically derived, but is also one of the earliest insecticidal products deployed in modern agriculture. Microbial formulations of Bt have been in use in pest control for almost 70 years. This lifespan of effectiveness is vastly longer than for typical chemical insecticides. Walker et al. (pp. 31-51) provide a review of current usage of Bt microbial formulations that includes killing of insect pests that harm a wide variety of crops, defoliate forests, and vector human diseases. Microbial Bt formulations are not as extensively used as many chemical insecticides, but still represent an important pest control option for farmers. This is especially so for farmers attempting to serve the market for 'organic' foods, where a litmus for acceptable production is whether the process can be identified as 'natural.' The long standing utility of Bt may be partly due to the complexity of the components conferring insecticidal activity, but is also undoubtedly linked to its limited use. In fact the only example of insects developing resistance to Bt in the field was linked to heavy use in Hawaii to control diamondback moth.

Use of Bt's insecticidal activity has recently become far less limited, appearing in a new form. With the advent of biotechnology-based crops, Bt was one of the first sources of material for the genetic engineering of new traits. Plants have been endowed with genes to produce the insecticidal crystal proteins (ICPs) that provide the primary insect-killing activity of Bt microbial formulations. Ingestion of part of the plant results in toxicity to a susceptible target pest. Different ICPs from various strains of Bt provide crops with protection from major Lepidopteran and Coleopteran pests. This is effective against pests that burrow into host plant tissues and have generally been hard to detect until damage has been done. Burrowing pests have also been difficult to target with many insecticides, including microbial Bt formulations, because the application is on the surface of the plant.

A great number of crops have been engineered to express Bt ICPs (see the article by Sharma et al., pp. 53-76, for a detailed account). Wide scale commercial deployment of Bt corn and cotton in the U.S. is the most prominent example, and these were some of the first genetically engineered crops to be marketed. Numerous concerns have encircled the release of Bt crops, some specific to genetic engineering, and others endemic to agriculture in general. Bt serves as a model for considering issues of modern agriculture and genetic engineering (Nester et al. 2002). Many of these issues are related to development, use and management of agricultural technology, and can be evaluated with the reports in this volume serving as a foundation.

Genetic engineering is often opposed as a device of profit-driven multinational corporations. Whether or not it is valid to uniformly vilify corporate agriculture is outside the scope of this work. However, it is clear (Sharma et al., pp. 53-76) that there is wide-scale interest in producing Bt crops with relevance to smallholding and poor farmers. In fact the number of regional staple crops, vegetables, and traditional cultivars that have been targets of genetic modification with Bt genes has surpassed the number of major commercial crops (primarily cotton and corn). More detailed accounts of efforts to generate Bt rice (see the article by Datta et al., pp. 77-91) and Bt potato (see the article by Ghislain et al., pp. 93-113) in varieties that are in use under the growing and cultural conditions of poor farmers in various regions of the developing world illustrate this targeted effort. These examples, and many other Bt crops (see Appendix) are part of public domain efforts to improve agriculture. Thus, genetic engineering of crops is at least as much a tool for the benefit of smallholding and poor farmers as it is a tool of corporate agriculture.

As with any pesticide, heavy use of Bt can provide selection pressure for the development of pests with resistance. For a detailed review of the mechanisms that can provide Bt resistance in pests, see Ferré and Van Rie (2002). The possibility of wide scale use of Bt crops selecting for resistant pests is a concern raised by critics of genetic engineering. It should also be a concern of any thoughtful proponent of use of the technology. Development of resistant pests would undermine any potential benefits of using Bt crops. Simply abstaining from using a technology will also obviate any potential benefits that it offers, so the most productive approach is appropriate management to counteract the development of Bt-resistant insects.

Many strategies are available to counteract development of Bt resistance in pests. A general description of numerous strategies is provided by Sharma et al. (pp. 53-76). One approach for combating resistance development is to stack the odds against resistance development by using multiple pesticides with different modes of action (gene pyramiding). This requires that pests simultaneously develop more than one resistance mechanism at the same time, which exponentially reduces the chances of resistance developing. Datta et al. (pp. 77-91) discuss efforts to introduce multiple Bt ICPs into rice for the purpose of delaying resistance development. Experiments in cauliflower by Cao et al. (pp. 193-207) demonstrate the viability of this strategy by showing that diamondback moth larvae resistant to Cry1A are still susceptible to Cry1C. The probabilistic advantage of multiple pesticidal components is of course only effective if introduced before resistance to one of the components has emerged.

Different technological and management strategies for combating resistance development may be appropriate and practical under different conditions. Matten and Reynolds (pp. 137-178) give a thorough account of U.S. EPA regulations for deployment of Bt cotton, where refuge planting is required, and monitoring and remedial action plans upon the discovery of resistant pest populations are provided. This management approach promises to be effective in cases where sufficient regulatory infrastructure, education, and farmer participation is present. In cases where these elements are lacking, built-in technological solutions may provide the most defense against failure of the plant protection. In addition to multiple pesticidal components, built-in options include approaches such as regulation of insecticide gene expression by tissue specific promoters, as is described in the research presented by Braga et al. (pp. 209-221) in sugarcane.

Genetically engineered crops have been portrayed as a force that will exacerbate the problem of monoculture in modern agriculture. Initially only a few traits were commercially released in a limited number of genetically engineered cultivars of a limited number of crops. However, this characterization misstates both the current situation and the future potential of the technology. Genetic engineering can introduce traits into different genetic backgrounds more rapidly than classical breeding.

An increasing number of crop species and cultivars are being successfully genetically engineered. Diversity of crop varieties will probably be promoted since the same amount of effort and time can introduce desirable modifications into more crops by genetic engineering than is possible through conventional breeding. Furthermore, traits that are not available through classical breeding can be added to the genetic makeup of crops through genetic engineering. Examples of genetic engineering directed at numerous cultivars of different crops are illustrated in articles four through six (pp. 53-113). Continued improvement of transformation technology will enable more and more crop cultivars to be practical targets of genetic engineering. Rajasekaran (pp. 179-192) describes a method for rapid evaluation of transgene constructs in numerous genetic backgrounds of cotton. Cao et al. (pp. 193-207) have developed methods for effectively performing *Agrobacterium*-mediated transformation of six different cauliflower cultivars. As stressed by Ghislain et al. (pp. 93-113), the genetic diversity of many crops may become neglected by farmers if they decide to use desirable engineered traits that are not available in traditional varieties. Thus, the ability to transfer the traits that farmers desire into many crop cultivars will be essential to providing farmers with the most options and incentives for conserving and using genetic diversity of crops.

Genetic engineering has the potential to facilitate avoidance of monoculture in agricultural systems, but does present potential problems where the engineered traits generate a monogenic feature in a cropping system. In the case of Bt, monogenic insecticidal protection in large areas of crop cultivation will generate strong selection pressure for development of pest resistance to the Bt ICP used. This could be especially problematic where a pest, such as cotton boll worm/corn ear worm, might encounter the same insecticide expressed in multiple crops that it feeds on. Strategies for combating resistance, which include use of a diverse set of transgenic traits, will be essential to the continued effectiveness of any transgenic pest protection of crops.

The environmental impact of Bt crops carries potential risks that require proper management, and benefits that deserve exploitation. Zipf and Rajasekaran (pp. 115-135) review the implications of Bt crops for the environment. In cases where sexually compatible wild relatives of a Bt crop are near a planting site, transgene flow into the wild relative should be considered. In some cases the barriers to successful crossing will make gene flow improbable, such as with potato (see the article by Ghislain et al., pp. 93-113), but in others gene flow will be an imminent possibility. If the introduced transgene provides a selective advantage, such as resistance to an insect that keeps the wild plants in check, then the transgene will be maintained in the population. A transgenic trait conferring a fitness advantage to a wild population could upset ecosystem dynamics and weed management in agricultural areas (Snow 2002). For this reason, a rigorous evaluation of the potential for flow of transgenes that could confer a selective advantage to wild plant populations is needed for any new crop/trait combination on a biogeographical scale. Technological and regulatory measures to prevent ecologically undesirable transgene flow are needed to assure that the benefits of Bt crops are sustainable.

Positive environmental impacts of Bt crops are one strong motivation for their adoption. The protection provided by Bt transgenes has been very effective in reducing chemical insecticide use. Zipf and Rajasekaran (pp. 115-135) provide an example of this, reviewing the impact of Bt cotton. The research presented by de Bianconi (pp. 223-235), comparing farming inputs between Bt and non-Bt cotton grown in Argentina, shows that farmers can dramatically reduce their insecticide use and maintain high yields with Bt cotton. The reduced input expenditures (insecticides and application costs) typically provide a substantial economic advantage to the growers of Bt crops in addition to improving environmental and worker safety.

Issues of economic control over crops and trade of agricultural products permeate the controversy over genetically engineered crops at many levels. This issue is more than simply about what parties are the primary intended beneficiaries of the technology. One high profile issue is the saving of seed between harvests by farmers. This activity is prohibited with private sector-owned genetically engineered crops, Bt and otherwise. But the barrier to seed-saving is not a trade-off introduced by genetic engineering. For many years farmers have turned to seed companies for hybrid seed. Yearly seed purchases, while creating interdependence of the two sectors, has enabled farmers to satisfy constant market pressure to obtain higher productivity and uniformity. Farmers in many areas have broadly opted for the advantages of hybrid seed over saved-seed, and in light of the impossibility of establishing their own breeding programs, have turned to seed companies to provide this product. Many of the Bt engineered crops generated in the public domain will not carry the impediment to seed-saving of corporate owned varieties. Where seed cannot be effectively saved, such as with the hybrid rice described by Datta et al. (pp. 77-91), seed-saving is problematic not by virtue of the biotechnology component, but by virtue of the heterozygosity of hybrid seed. Genetic engineering cannot be justly characterized as a vehicle for undermining traditional seed-saving practices, because in instances where engineered seed cannot be saved, markets have already largely transitioned to a practice of obtaining seed from a supplier on a yearly basis.

International agricultural trade is another complex economic issue that has been contentious for genetically engineered crops. Wide variation in public perceptions, regulatory frameworks, and trade interests has complicated the marketing of these crops. In fact, at times it has even proven difficult to give away genetically engineered food. At the end of 2002 the Zambian government decided to reject corn food aid from the U.S. despite an impending famine. The justification given for this decision confounds logic: concerns cited were that Zambia might lose European countries as potential markets for grain exports, and that the long-term health effects of consuming genetically engineered food are not clearly understood. Zambia becoming a food exporting country is probably at least as distant as acceptance of genetically engineered foods in Europe. The hypothetical risks of eating genetically engineered foods, part of the mainstream U.S. market since 1996, pale in comparison with the reality of death by starvation.

Labeling requirements for genetically engineered foods can produce challenges for growers and processors, introducing new segregation costs and potential stigmatization in some markets. As pointed out by Ghislain et al. (pp. 93-113), labeling requirements can be expected to complicate the ability of poor and smallholding farmers to reap the benefits of growing Bt and other genetically engineered crops. It is ironic that the movement to label and marginalize genetically engineered crops states one of its motivations to be opposition to a tool that could be advantageous to corporate farming. Labeling requirements will likely exclude poor and smallholding farmers from being able to fully access the benefits of growing genetically engineered crops, being competitive, and improving their livelihoods. The cost of wholesale labeling of genetically engineered crops will be large, perhaps prohibitive for many farmers and processors, while the benefits to consumers are uncertain.

Many political, economic, and social phenomena have contributed to how genetic engineering is used and not used. The technical focus of this project on development, use and management of Bt crops does not provide for in-depth consideration of these non-technical issues. However, these issues will continue to influence the trajectory that this technology takes in delivering benefits to society, and how potential hazards are met. A detailed review of these issues has recently been produced by Shelton et al. (2002). Bt has been and will likely continue to be at the forefront of innovations for, and deliberations about modern agriculture.

Matthew Metz
AAAS Diplomacy Fellow
U.S. Agency for International Development
1300 Pennsylvania Avenue, NW
Washington, DC 20532-2110
E-mail: mmetz@usaid.gov

REFERENCES

Ferré, J. and J. Van Rie (2002) Biochemistry and Genetics of Insect Resistance to *Bacillus thuringiensis*. Annual Reviews of Entomology 47:501-33.

Nester, E. W., L. Thomashow, M. Metz and M. Gordon (2002) 100 Years of *Bacillus thurnigiensis*–a critical scientific assessment. American Academy of Microbiology.

Shelton, A. M., J.-Z. Zhao and R. T. Roush (2002) Economic, Ecological, Food Safety, and Social Consequences of the Deployment of Bt Transgenic Plants. Annual Reviews of Entomology 47:845-81.

Snow, A. A. (2002) Commentary: Transgenic Crops–Why Gene Flow Matters. Nature Biotechnology 20:542.

APPENDIX

Examples of crops engineered with Bt		
Apple	Cotton	Rice
Broccoli	Eggplant	Sorghum
Cauliflower	Groundnut/Peanut	Sugarcane
Chickpea	Pigeonpea	Tobacco
Corn	Potato	Tomato

The Safety of *Bacillus thuringiensis* for Human Consumption

John Kough

SUMMARY. *Bacillus thuringiensis* (Bt) provides one of the most prominent biologically based pesticides. Bt affords a high degree of safety to the environment, non-target organisms and humans in combination with its effectiveness in controlling numerous insect pests. These features fill a pressing need to produce food and fiber for a growing population without eroding environment and health. Extensive safety testing and a history of human exposure without adverse effects have proven Bt to be one of the most important reduced risk pesticides. Not only is Bt effective against targeted pests, but it is safe enough to humans to be exempted from many of the precautions that apply to conventional synthetic chemical pesticides. *[Article copies available for a fee from The Haworth Document Delivery Service: 1-800-HAWORTH. E-mail address: <docdelivery@haworthpress. com> Website: <http://www.HaworthPress.com>]*

KEYWORDS. *Bacillus thuringiensis* (Bt), *B. cereus*, toxicity, insecticidal crystal protein (ICP), hypersensitivity

John Kough is affiliated with the United States Environmental Protection Agency, Office of Pesticide Programs, Biopesticides and Pollution Prevention Division (7511C), 1200 Pennsylvania Avenue, NW, Washington, DC 20460.

The views expressed in this chapter are those of the author and do not necessarily represent those of the United States Environmental Protection Agency (EPA). The use of trade, firm, or corporation names in this article is for the information and convenience of the reader. Such use does not constitute an official endorsement or approval by EPA of any product or service to the exclusion of others that may be suitable.

[Haworth co-indexing entry note]: "The Safety of *Bacillus thuringiensis* for Human Consumption." Kough, John. Co-published simultaneously in *Journal of New Seeds* (Food Products Press, an imprint of The Haworth Press, Inc.) Vol. 5, No. 1, 2003, pp. 1-10; and: *Bacillus thuringiensis: A Cornerstone of Modern Agriculture* (ed: Matthew Metz) Food Products Press, an imprint of The Haworth Press, Inc., 2003, pp. 1-10. Single or multiple copies of this article are available for a fee from The Haworth Document Delivery Service [1-800-HAWORTH, 9:00 a.m. - 5:00 p.m. (EST). E-mail address: docdelivery@haworthpress.com].

INTRODUCTION

Demands for increased food and fiber production from a finite amount of arable land and a heightened awareness of environmental impact from pest control measures have made reduced risk pesticides critical. These pesticides include biologically based active ingredients such as insect pheromones and viruses, microbial antagonists for plant diseases, inducers of systemic acquired resistance and microbial pathogens of insects like *Bacillus thuringiensis* (Bt). One of the major functions of the Environmental Protection Agency's Office of Pesticide Programs (OPP) under its authorities from the Federal Insecticide, Fungicide and Rodenticide Act is to insure that pesticide use does not present an unreasonable adverse effect to the environment. OPP is also responsible under the Federal Food Drug and Cosmetic Act to determine that the aggregate exposure to pesticide residues, including reduced risk biopesticides, has a reasonable certainty of causing no harm. OPP does this by reviewing data on the safety of these products and registering the pesticide with an approved label describing its correct use. For most microbial biopesticides, notably Bt, the safety data also justifies an exemption from the requirement of a food tolerance (i.e., an upper limit on acceptable pesticide residues on a food crop).

The exemption from a food tolerance in essence indicates that a pesticide, applied according to the approved label, does not present a significant dietary risk if the treated commodity is consumed shortly after product application. Most conventional synthetic chemical pesticides have strict requirements that the pesticide cannot be applied for a specified period prior to harvest. This is termed a pre-harvest interval and is intended to allow the pesticide residues to weather and degrade in the environment to levels that are safe to consume. The granting of a food tolerance exemption to a product provides the advantage of more flexibility for use on food crops.

Many microbial pesticide active ingredients utilizing *Bacillus thuringiensis* (Bt) strains have been registered (see Walker et al., this volume, pp. 31-51) since the first Bt product was granted U.S. approval in 1961. Bt formulations have become a major pest control option for use on vegetable crops due to their effectiveness and lack of a pre-harvest interval. The companies that have registered these Bt products have all submitted data used to determine their safety for human health and use in the environment. In order to better understand the dietary safety assessment for Bt products, it is necessary to discuss the data requirements that must be supplied for registration of a microbial pesticide. The information used for a safety determination includes the results of tests done with surrogate mammalian species, and specific details of the taxonomic identification and biology of the microbe in question. The latter data, termed product characterization, provide a scientific foundation to examine effects that might be evident in the toxicity tests as well as specific issues relating to the manufacturing process. The product characterization data specifically define the microbe that is the active pesticidal ingredient, and provide a historical background to judge the extent of prior human exposure and any reported adverse effects associated with the microbe.

Unlike conventional synthetic chemical pesticides, microbial biopesticides such as Bt are living organisms. The identification of the microbe, therefore, is quite different from the analytical methods used to describe a conventional chemical pesticide. The method of manufacture also demands different quality controls than are typical of a chemical synthesis scheme. A special set of safety tests designed to address the hazards associated with infectivity and pathogenicity have been utilized in order to determine the safety of using these biological agents. The guidelines describing how the toxicity studies should be done as well as details for microbial identification and the manufacturing process description are found at the following website: (http://www.epa.gov/docs/ OPPTS_Harmonized/885_Microbial_Pesticide_Test_Guidelines/Series/).

PRODUCT CHARACTERIZATION

The identification of Bt microbes has become more complicated across its history. A bacterial disease of silkworms (*Bombyx mori*) was first recognized by the Japanese bacteriologist Ishiwata in 1902, and the causal bacterium was formally named *Bacillus thuringiensis* by Ernst Berliner in 1915 (Beegle, 1992). Bt isolates have generally been described as Gram-positive aerobic, or facultative anaerobic rods having a crystalline inclusion body and found in sporulating colonies with pathogenicity to insects. Since those original descriptions, much taxonomic and biochemical work has been done to distinguish among Bt isolates and further understand the nature of insect pathogenicity. The importance of the parasporal crystal inclusion (Cry) proteins in insect pathogenicity and bacterial classification has been established (Sneath, 1986; Schnepf, 1998). They are also referred to as ICPs, insecticidal crystal proteins. The range of animals susceptible to intoxication by pure ICPs has expanded to include not only insect species in the Lepidoptera,

Coleoptera and Diptera families but also to nematode species (Crickmore, 1998). The mode of action of the ICPs in insect mortality has been established as insect midgut membrane binding and disruption (English, 1992). The binding specificity of ICPs for insect gut receptors and not mammalian species has also been indicated (Noteborn, 1995). Exploitation of the ICPs themselves is currently the focus for genetically engineered plants. There is an additional insecticidal effect of the septicemia ascribed to the presence of the bacterium itself, thus in the assessment of the Bt products for mammalian safety, many of the areas of concern relate to the Bt bacterium itself.

The biology of taxonomically related organisms is an important consideration in the description of a microbial pesticide. The other species within the genus *Bacillus* that are most closely related to Bt are *B. cereus*, *B. cereus* var. *mycoides* and *B. anthracis* (Sneath, 1986; Guttmann, 2000). It is required that any new strain of Bt seeking registration as a biopesticide be adequately characterized to distinguish it from these related species using techniques such as biochemical capabilities, serotyping and bioactivity. It is also important that the quality control in manufacture of any Bt product insure that these related species, as well as any human pathogenic microbes, are not present in any product at levels that could be a hazard to humans.

The fact that *B. cereus* is nearly identical to Bt in most biochemical characteristics except for the presence of the hallmark parasporal crystal inclusion body in Bt has lead several authors to the conclusion that Bt is simply a *B. cereus* capable of forming parasporal inclusion bodies (Carlson, 1994; Zahner, 1989). Since the majority of ICPs expressed in the examined Bt strains are borne on plasmids potentially capable of being moved between related bacteria, this supposition has certain merit (Gonzalez, 1982). *B. cereus* strains with the ability to produce protein enterotoxins have been shown to be the causal agent in incidents of food poisoning (USFDA, 1992; Turnbull, 1981). Enterotoxins similar to those in *B. cereus* have been suggested to also be present in Bt (Damgaard, 1995). Some Bt isolates produce β-exotoxin, a heat stable toxin that is an adenine nucleotide analogue and has demonstrated mammalian toxicity (de Barjac, 1969; Vankova, 1978). There have also been reports of Bt being isolated from clinical specimens (Hernandez, 1998; Samples, 1983). Surveys of naturally occurring isolates of Bt expressing one type of ICP have also found that environmental isolates of Bt display significant variation in the expression of the other toxins that have mammalian effects (Perani, 1998). All these factors make the best taxonomic identification and screening for known or suspected bacterial toxins of new Bt isolates crucial aspects in the safety assessment. The success of these assessments is indicated in that the currently registered strains have no indication of presenting any of the hazards discussed above (McClintock, 1995a; USEPA, 1998).

MANUFACTURING PROCESS

The taxonomic data and biology provide a historical background on human exposure, potential confounding microbial contaminants and suspect components potentially made by the active microbial ingredient. The manufacturing process can help address many of these issues. For Bt products, like all pesticides, a description of the fermentation process is provided to EPA during the registration evaluation. This information is considered confidential business information and cannot be discussed in other than general terms in public documents. However, every manufacturing process description should address detection and control of unwanted microorganisms and contamination by suspected toxic components. Particular attention is given to the measures to minimize the potential growth of contaminating organisms.

The quality assurance/quality control (QA/QC) procedures used to insure a uniform Bt product include: (1) proper maintenance of stock and "seed" cultures used to begin the fermentation of a microbial agent as well as analyses for biological purity; (2) a description of sterilization procedures for the growth media and fermentation vessels; (3) monitoring of appropriate parameters for the physical conditions during fermentation (e.g., O_2, CO_2, pH), and (4) QA/QC analysis of lots at critical steps and when fermentation is completed. The QA/QC techniques also usually include a bioassay against a target pest or a quantification of ICPs present for product acceptance. This information, along with the results of analyses from five production batches, provides a framework to examine the adequacy of the QA/QC measures to control toxic or sensitizing materials arising from growth of contaminating microorganisms in the pesticide product. QA/QC procedures that control or remove toxic or sensitizing ingredients are also essential if the Bt has been shown to have the potential to produce compounds like β-exotoxin. In the tolerance exemption for Bt products (40 CFR 180.1011), it is specified that each "master seed lot" be screened for the isolate's ability to produce β-exotoxin or, if appropriate, production batches are periodically examined for the presence of β-exotoxin to insure that manufacturing procedures eliminate it from the final product.

If the production method can support growth of human or animal pathogens, each batch of a microbial pesticide is analyzed for the presence of human or animal pathogens (e.g., *Shigella*, *Salmonella*, and *Vibrio* or an indicator organism). The manufacturing process must also include methodologies for elimination of these pathogens from the production batch if contaminated batches are not discarded. For Bt fermentation batches, each lot is also tested ". . . by subcutaneous injection of at least 1 million spores into each of five laboratory test mice." The test results must show ". . . no evidence of infection or injury in the test animals when observed for 7 days following injection" (40 CFR 180.1011). This added quality control step ensures that there is no possibility of *B. anthracis* or other significant human pathogens being released in a Bt product.

TOXICITY/INFECTIVITY TESTING

The mammalian toxicology data requirements are structured in a tiered testing system. The focus is on those studies necessary for a human health risk assessment providing a reasonable certainty of no harm from the aggregate exposure to the pesticide residues. Registration of a Bt microbial pesticide for use on a terrestrial food crop requires studies that include acute infectivity/toxicity tests with the technical grade active ingredient. These tests examine oral, pulmonary and injection exposures. After dosing, test animals are evaluated by recording mortality, body weight gain and clinical signs of toxicity. In addition, the test animals are examined by performing a gross necropsy, and evaluating the pattern of clearance of the microorganism from the animals. For establishing a pattern of clearance, the Bt microorganism is periodically enumerated from appropriate organs, tissues, and body fluids of test animals to verify the lack of pathogenicity/infectivity or persistence in mammals and to document normal immunological processing of the Bt inoculum. The only unusual findings of submitted infectivity tests was a prolonged pattern of clearance, most probably due to the Bt being present as spores in the test substance (McClintock, 1995b; EPA, 1998). In addition, tests on the toxicity and irritation of the formulated end-use product containing Bt have provided toxicity categories for use on the product label. To date the end-use product toxicity/irritation tests of these Bt products have indicated very slight to practically no significant toxicity.

Hypersensitivity (i.e., dermal sensitization) studies are generally not required for registration of microbial pesticide products; injection induction and challenge with microbial pesticides, including protein-aceous components, into the commonly used laboratory animal (i.e., guinea pig) would be expected to reveal any potential interaction. This, coupled with a history of experience with fermentation products have allowed for the conclusion that allergic responses to microbial pesti-cides should have been observed during development and manufacture if there exists significant risk for dermal sensitization. However, once a Bt product is registered, the company must submit to EPA any informa-tion/data on incidents of hypersensitivity, including immediate-type and delayed-type reactions of humans or domestic animals that occur during the production and use of the technical grade of the active ingre-dient, the manufacturing-use product, or the end-use product. The re-quirements for adverse incident reporting in connection with use of a registered Bt product can be found in section 6(a)(2) of the Federal In-secticide, Fungicide and Rodenticide Act and 40 CFR Part 159.

While no reports of adverse effects submitted to EPA relating to Bt products have been verified as being due to the Bt active ingredient, there has been a literature report about potential sensitization of farm workers (Bernstein, 1999). This study shows that farm workers exposed to crops treated with Bt products can be shown to have circulating immunoglobulins to both spore and vegetative forms of Bt. All workers showed a level of immunoglobulin response, with more highly exposed workers showing a higher response. While there were observed immu-noglobulin E responses and positive skin prick tests, there were no clini-cal symptoms of hypersensitivity. Examination of the components of the Bt extracts used to elicit the skin reaction also demonstrated that the ICP containing fraction was not responsible for the effect.

It is important to note that the identification of Bt isolates, the toxi-cology/infectivity test results, as well as the fermentation methods and QC procedures all indicate that there are no significant hazards associ-ated with the Bt products currently used in US agriculture. Other issues or data related specifically to the registration of *B. thuringiensis*, for ex-ample the environmental effects risk assessment, are discussed in the Reregistration Eligibility Document for *Bacillus thuringiensis* (USEPA, 1998).

CONCLUSION AND FUTURE DEVELOPMENTS

The information from the acute toxicity/pathogenicity studies on Bt strains has supported the assessment that the Bt microorganisms are not pathogenic to, or toxic to, mammals. The lack of adverse effects in toxicity tests for entire preparations suggests that the protein components of the microorganism and fermentation residues are not toxic to mammals. It is reasonable to conclude from these findings that the ICPs, a major source of the insecticidal activity in Bt preparations, have not been implicated in mammalian toxicity.

This history of mammalian toxicity tests using numerous Bt strains as well as the continued low incidence of reported adverse effects with use of microbial Bt products have been a significant aspect of the safety finding for genetically engineered plants expressing ICPs. In addition, significant tests examining the safety of isolated ICPs are consistent with the finding of a reasonable certainty of no harm for the aggregate exposure for the Bt engineered plants. The recent assessment of these products is available at the following website: (http://www.epa.gov/pesticides/biopesticides/reds/brad_bt_pip2.htm).

The discovery of new ICPs and other toxins expressed by Bt will undoubtedly lead to the introduction of new microbial pesticides. The fact that there is a broader range of insects infected by Bt than there are insects susceptible to the currently characterized ICPs implies that there is still a great unexploited potential of safe insecticidal products to be developed from the use of Bt strains.

REFERENCES

Beegle, C.G. and T. Yamamoto (1992) History of *Bacillus thuringiensis* Berliner Research and Development. Can. Ent. 124:587-616.

Bernstein, I.L. J.A. Bernstein, M. Miller, S. Tierzieva, D.I. Bernstein, Z. Lummus, M.J.K. Selgrade, D.L. Doerfler and V.L. Seligy (1999) Immune responses in farm workers after exposure to *Bacillus thuringiensis* pesticides. Environ. Health Perspect. 107:575-582.

Carlson, C.R., D.A. Caugant and A.B. Kolsto (1994) Genotypic diversity among *Bacillus cereus* and *Bacillus thuringiensis* strains. Appl. Environ Microbiol. 60: 1719-1725.

Crickmore, N., D.R. Zeigler, E. Schnepf, J. Van Rie, D. Lereclus, J. Baum and D.H. Dean (1998) Revision of the nomenclature for the *Bacillus thuringiensis* pesticidal crystal proteins. Microbiol and Molec. Biol. Rev. 62:807-813.

Damgaard, P.H. (1995) Diarrhoeal enterotoxin production by strains of *Bacillus thuringiensis* isolated from commercial *Bacillus thuringiensis*-based insecticides. FEMS Immunol. Medical. Microbiol. 12:245-256.

De Barjac, H. and J.Y. Riou (1969) Action de la toxine thermostable de *Bacillus thuringiensis* var. *thuringiensis* administree a des souris. Review of Pathological Comp. Medicine Experiment 6:19-26.

English, L. and S.L. Slatin (1992) Mode of action of delta-endotoxins from *Bacillus thuringiensis*: A comparion with other bacterial toxins. Insect Biochem. Molec. Biol. 22:1-7.

Gonzalez, J.M., B.J. Brown and B.C. Carlton (1982) Transfer of *Bacillus thuringiensis* plasmids coding for β-endotoxins among strains of *B. thuringiensis* and *B. cereus*. Proc. Natl. Acad. Sci. USA 79:6951-6955.

Guttmann, D.M. and D.J. Ellar (2000) Phenotypic and genotypic comparisons of 23 strains from the *Bacillus cereus* complex for a selection of known and putative *B. thuringiensis* virulence factors. FEMS Microbiol. Ltrs. 188:7-13.

Hernandez, E., F. Ramisse, T. Cruel, J.P. Ducoureau, J.M. Alonso and J.D. Cavallo (1998) *Bacillus thuringiensis* serovar H34-*konkukian* superinfection: report of one case and experimental evidence of pathogenicity in immunosuppressed mice. J. Clin. Microbiol. 36:2138-2139.

McClintock, J.T., C.R. Schaffer, J.L. Kough and R.D. Sjoblad (1995a) Relevant taxonomic considerations for regulation of *Bacillus thuringiensis*-based pesticides by the U.S. Environmental Protection Agency. In: T-Y Feng et al. (eds.), "*Bacillus thuringiensis* Biotechnology and Environmental Benefits," Vol. I, 313-325.

McClintock, J.T., C.R. Schaffer and R.D. Sjoblad (1995b) A Comparative review of the mammalian toxicity of *Bacillus thuringiensis*-based pesticides. Pestic. Sci. 45:95-105.

Noteborn, H.P.J.M., M.E. Bienenmann-Ploum, J.H.J. van den Berg, G.M. Alink, L/Zolla, A. Reynaerts, M. Pensa and H.A. Kuiper (1995) Safety assessment of *bacillus thuringiensis* insecticidal crystal protein CRYIA(b) expressed in transgenic tomato, pp. 134-147, In: *Genetically Modified Foods: Safety Issues*. American Chemical Society, Washington, DC.

Perani, M., A.H. Bishop and A. Vaid (1998) Prevalance of β-exotoxin, diarrhoeal toxin and specific β-endotoxin in natural isolates of *Bacillus thuringiensis*. FEMS Microbiol. Ltrs. 160:55-60.

Samples, J.R. and H. Buettner (1983) Corneal ulcer caused by a biologic insecticide (*Bacillus thuringiensis*). Am. J. Ophthalmology 95:258-260.

Schnepf, E., N. Crickmore, J. Van Rie, D. Lereclus, J. Baum, J. Feitelson, D.R. Zeigler and D.H. Dean (1998) *Bacillus thuringiensis* and its pesticidal crystal proteins. Microbiol. and Mol. Biol. Rev. 62:775-800.

Sneath P.H.A. (1986) "Endospore-Forming Gram-Positive Rods and Cocci," Section 13, pp. 1104-1139, In: Bergey's Manual of Systematic Bacteriology Vol. 2, P.H.A. Sneath et al. (eds.). Williams and Wilkins Publishers Baltimore, London, Los Angeles, Sydney.

Turnbull, P.C.B. (1981) *Bacillus cereus* toxins. Pharmac. Ther. 13:453-505.

U.S. Environmental Protection Agency, Office of Prevention, Pesticides and Toxic Substances (1998) Reregistration eligibility decision for *Bacillus thuringiensis*. (*http://www.epa.gov/oppsrrd1/REDs/0247.pdf*).

U.S. Food and Drug Administration, Center for Food Safety and Applied Nutrition (1992) "*Bacillus cereus* and other *Bacillus* species," Chapter 12, In: *Foodborne Pathogenic Microorganisms and Natural Toxins Handbook.* (http://www.cfsan.fda.gov/~mow/chap12.html).

Vankova, J. (1978) The heat-stable exotoxin of *Bacillus thuringiensis*. Folia Microbiologia 23:162-174.

Zahner, V., H. Momen, C.A. Salles and L. Rabinovitch (1989) A comparative study of enzyme variation in *Bacillus cereus* and *Bacillus thuringiensis*. J. Appl. Bacteriol. 67:275-282.

Effects of Bt on Non-Target Organisms

Brian A. Federici

SUMMARY. The development of Bt crops is one of the most significant advances in crop protection technology of the past fifty years. Current Bt crops are based on highly specific insecticidal Cry proteins of *Bacillus thuringiensis*. Foliar Bt insecticides based on the same proteins have been used for more than forty years and have a remarkable safety record, with no known detrimental effects reported on vertebrate or non-target invertebrate populations. Bt cotton and Bt corn have been widely adopted by farmers in the United States, with acreage averaging 40-50% of the area planted with cotton or corn in 2002. Concern has been raised about the safety of Bt crops. However, most evidence shows that these crops, like foliar Bt insecticides, are safe for non-target organisms, especially in comparison to chemical insecticides. Evidence for safety comes from knowledge of Cry protein mode of action as well as from studies of the effects of Cry proteins on non-target organisms tested in the laboratory and under operational growing conditions. These studies indicate that Bt crops, owing to their high degree of safety, can serve as the cornerstone for more environmentally sound integrated pest management programs. *[Article copies available for a fee from The Haworth Document Delivery Service: 1-800-HAWORTH. E-mail address: <docdelivery@ haworthpress.com> Website: <http://www.HaworthPress.com> © 2003 by The Haworth Press, Inc. All rights reserved.]*

Brian A. Federici is Professor of Entomology and Member, Interdepartmental Graduate Programs in Genetics and Microbiology, University of California-Riverside, Riverside, CA 92521 (E-1

[Haworth co-indexing entry note]: "Effects of Bt on Non-Target Organisms." Federici, Brian A. Co-published simultaneously in *Journal of New Seeds* (Food Products Press, an imprint of The Haworth Press, Inc.) Vol. 5, No. 1, 2003, pp. 11-30; and: *Bacillus thuringiensis: A Cornerstone of Modern Agriculture* (ed: Matthew Metz) Food Products Press, an imprint of The Haworth Press, Inc., 2003, pp. 11-30. Single or multiple copies of this article are available for a fee from The Haworth Document Delivery Service [1-800-HAWORTH, 9:00 a.m. - 5:00 p.m. (EST). E-mail address: docdelivery@haworthpress.com].

http://www.haworthpress.com/store/product.asp?sku=J153
© 2003 by Taylor & Francis.
10.1300/J153v05n01_02

11

KEYWORDS. *Bacillus thuringiensis*, transgenic crops, non-target organism, Bt cotton, Bt corn, Cry proteins

INTRODUCTION

Since World War II, synthetic chemical insecticides have been the most extensively employed agents for controlling the wide range of insect pests that attack crops. Moreover, hundreds of millions of pounds of active ingredients of these chemicals are still used annually throughout the world. While there is little doubt that these chemicals enabled production of a bountiful food supply for most countries, their detrimental effects on non-target vertebrate and invertebrate populations has been recognized for decades. Moreover, the public is more concerned than ever about the effects of synthetic chemical insecticides on their own health, as is evident from increased sales of organic foods.

Chemical insecticides have become more specific as well as more biodegradable decade by decade, yet most still have a spectrum of activity that is quite broad, generally affecting most members of multiple orders of insects. The increased specificity of new chemical insecticides is a positive development, yet by far the most significant advance of the last 50 years in decreasing the adverse effects of chemical insecticides is the development of insecticidal transgenic crops based on the Cry proteins of *Bacillus thuringiensis* (Bt). More than 40 million acres of these crops, referred to as Bt crops, are now planted annually in the United States. This acreage consists of primarily of Bt corn and Bt cotton used to control caterpillar pests such as the European corn borer, *Ostrinia nubilalis*, the pink bollworm, *Pectinophora gossypiella*, and species of *Heliothis* and *Helicoverpa*. Bt corn active against the corn rootworm, *Diabrotica undecimpunctata*, is now registered, further increasing the acreage of Bt crops in the United States.

Bt crops have contributed to major reductions in pesticide derived environmental impact due to the high degree of insecticidal specificity of these crops. However, a few laboratory studies, especially one showing that Bt corn pollen could kill larvae of the monarch butterfly, quickly gained media attention and led to widespread concern in the scientific community and the public about the safety of these crops to non-target organisms. Despite the political danger, the U.S. Environmental Protection Agency has maintained standards of using the results of experimental studies and risk-assessment procedures to determine the safety of Bt crops to non-target vertebrate and invertebrate organisms, including humans. Based on these studies, the U.S. EPA has allowed existing registrations to remain in effect, and is proceeding with evaluation of petitions to register new insecticidal transgenic crops incorporating Bt proteins. Nevertheless, concern remains that Bt crops may be detrimental to non-target organisms such as pest natural enemy populations and endangered insect species. Thus, the purpose of this paper is to review the data emerging from a variety of studies on the effects Bt Cry proteins used in Bt cotton and Bt corn on non-target organisms. Laboratory studies are not always representative of effects likely to occur under actual field conditions. The focus of this brief overview, therefore, is first on the scientific basis for the safety of Bt to non-target organisms, followed by review of key data from previous and ongoing non-target studies carried out under laboratory and field conditions. These studies indicate Bt crops will prove of significant benefit to the environment as well as to human and animal health.

BASIC BIOLOGY OF BACILLUS THURINGIENSIS

Bt is a common gram-positive, spore-forming bacterium that can be readily isolated on simple media (i.e., nutrient agar) from a variety of habitats including soil, water, plants, grain dust, dead insects, and insect feces. Its life cycle is simple. When nutrients are sufficient for growth, the spore germinates producing a vegetative cell that grows and reproduces by binary fission. The bacterium continues to multiply until one or more nutrients, such as sugars, certain amino acids, and oxygen, become insufficient for continued vegetative growth. Under these conditions, the bacterium sporulates producing a spore and parasporal body, the latter composed primarily of insecticidal proteins in the form of crystalline inclusions (Federici, 1999). These are commonly referred to in the literature as protein toxins or endotoxins, with recent nomenclature favoring the term insecticidal crystal proteins (ICPs). There are two basic types of ICPs, Cry proteins, Cry standing for "crystal," and Cyt proteins, Cyt standing for "cytolytic." The latter are not used in transgenic plants and will not be discussed here. The Cry proteins used in transgenic plants consist of three structural domains: Domain I, a pore-forming domain; Domain II, a membrane-binding domain; and Domain III, responsible for stability and membrane binding (Li et al., 1991). Current Bt crops use Cry proteins, of which there are three major insect specificity types, those active against either: (1) Lepidopetera (butter flies/moths); (2) Diptera (mosquito/fly) larvae; or (3) Coleoptera (beetles). It is important to realize that there are no Bt proteins active against all three of these groups at a level considered of commercial utility. The specific toxicity of several major Bt Cry proteins against a range of insects is shown in Table 1.

MODE OF ACTION OF CRY PROTEINS

Cry proteins are actually protoxins that must be ingested and processed by enzymes to yield an active toxin (Schnepf et al., 1998). Knowledge of Cry protein mode of action, as summarized below, in combination with the data in Table 1, provides a basis for understanding the specificity and the resultant safety of Bt insecticide formulations and Bt crops. The different levels of specificity can be defined as follows:

1. Endotoxin crystals must be ingested to have an effect; there is no "contact" activity, as occurs with chemical insecticides. This is one reason sucking insects and other invertebrates such as spiders and mites are not sensitive to Bt (Federici, 1999).
2. After ingestion, Bt endotoxin crystals with potential toxicity against lepidopterous insects must be activated to be toxic (Schnepf et al., 1998). A prerequisite for activation is that crystals dissolve. This requires alkaline conditions, generally a midgut pH in the range of 8 or higher. Most non-target invertebrates have neutral or only slightly acidic or basic midguts. Under the highly acidic conditions in stomachs of many vertebrates, including humans, endotoxin crystals may dissolve, but the solubilized proteins are rapidly degraded by gastric juices to non-toxic peptides, typically within minutes.
3. Once solubilized, activation occurs when Cry proteins are cleaved by midgut proteases at both the c-terminus and n-terminus (Bravo et al., 2002).
4. Once activated, the toxin must bind to glycoprotein "receptors" on midgut microvillar membrane (Schnepf et al., 1998). Most chewing insects that ingest toxin crystals, even those with alkaline midguts, including many lepidopterans, do not have the appropriate receptors, and thus are not sensitive to activated Cry proteins. Even insects sensitive to one class of Bt proteins, such as lepidopterans sensitive to Cry1 proteins, are not sensitive to Cry3A active against coleopterans, as they lack receptors for these. Moreover,

no binding of Cry protein has been detected in mammalian stomach epithelial cells (Noteborn, 1995).

5. After binding to a midgut receptor, the toxin must enter the cell membrane, change conformation in the process, and oligomerize to form pores that will be toxic (Knowles & Dow, 1993).

With respect to level 5, the specific conformational changes that must take place to exert toxicity are not currently known. It is known, however, that high affinity irreversible binding can occur in some insects, yet not lead to toxicity. This implies that a specific type of processing, i.e., another level of specificity, is required for toxicity that occurs as or after the toxin inserts into the membrane.

In Bt crops, a portion of the second level (i. e., level 2) of the first five levels of specificity has been circumvented. When synthesized in plants, full length and truncated Cry proteins do not form crystals, and even if quasi-crystalline inclusions do form, toxin remains in solution within the plant cells. Nevertheless, whether produced in plants as a full length or truncated protoxin, Cry proteins must still be properly activated after ingestion, i.e., cleaved properly at the c and n termini. Additionally, they must meet the criteria for binding and membrane insertion defined above by levels 4 and 5 to be toxic. Furthermore, with the one exception Cry9C, which was engineered to resist rapid proteolytic cleavage, most Bt proteins produced in Bt crops are degraded rapidly under conditions that mimic the mammalian digestive system. Thus, most of the inherent levels of specificity that account for the safety of Cry proteins used in

TABLE 1. Toxicity of Bt Cry Proteins to Lepidopteran and Dipteran Pests*

| Cry Protein | LC$_{50}$ in ng/cm^2 of diet or water** | | | | |
	Tobacco Hornworm	Tobacco Budworm	Cotton Leafworm	Yellow Fever Mosquito	Colorado Potato Beetle
Cry1Aa	5.2	90	> 1,350	> 5,000	> 5,000
Cry1Ab	8.6	10	> 1,350	> 5,000	> 5,000
Cry1Ac	5.3	1.6	> 1,350	> 5,000	> 5,000
Cry1C	> 128	> 256	104	> 5,000	> 5,000
Cry11A	> 5,000	> 5,000	> 5,000	60	> 5,000
Cry3A	>5,000	>5,000	> 5,000	> 5,000	< 200

*Toxicity to first instars of the Tobacco Hornworm (*Manduca sexta*), Tobacco Budworm (*Heliothis virescens*), Cotton Leafworm (*Spodoptera littoralis*), Yellow Fever Mosquito (*Aedes aegypti*), Colorado Potato Beetle (*Leptinotarsa decimlineata*). Modified from Hofte and Whiteley (1989).
**Values of > 5,000 indicate a lack of toxicity at high doses, that is doses that are equivalent to field applications rates that would not be economical. Lack of toxicity at these rates, however, also illustrates the high degree of insect specificity demonstrated by Cry endotoxins.

commercial bacterial insecticides apply to these same proteins when used to make Bt crops resistant to insects.

Lastly, an important concept of evaluating safety is to consider the route by which an organism is likely to encounter a toxin. Even though pulmonary (inhalation) and intraperitoneal injection studies are done with microbial Bt insecticides and proteins, their normal route of entry by target and non-target organisms is by ingestion. This is likewise true for Bt proteins produced in Bt crops.

SAFETY TESTING OF Bt

In addition to their insecticidal efficacy, a major impetus for using Cry proteins in Bt crops was their long history of safety to non-target organisms, especially to vertebrates. The most important levels of Bt ICP specificity described above, i.e., activation, binding, and membrane insertion, apply equally to evaluating the safety of Cry proteins whether used in Bt crops or bacterial insecticides. Therefore, the tests and data that support a very high degree of safety for microbial insecticide formulations containing Cry proteins are relevant to assessing the safety of Bt crops (Kough, this volume, pp. 1-10).

Extensive testing has been and remains required to meet rigorous safety requirements established by governmental agencies such as the U.S. Environmental Protection Agency. Use of data from these tests is valid as the major approach to evaluate Bt crop safety, especially considering that many hundreds of safety tests have been conducted over several decades to register numerous microbial insecticide formulations based on different subspecies of Bt (see Walker et al., this volume, pp. 31-51). The principal Bt subspecies evaluated in these tests have been *B. t.* subsp. *kurstaki*, *B. t.* subsp. *israelensis*, *B. t.* subsp. *aizawai*, and *B. t.* subsp. *morrisoni* (strain tenebrionis). The materials evaluated have been the active ingredients, i.e., sporulated cultures containing spores and crystals of Cry and Cyt proteins, as well as formulated products. Among the materials tested are all of the Cry proteins used in commercial Bt crops currently on the market.

In determining what types of tests should be done to evaluate the safety of bacterial insecticides, early tests were based primarily on those used to evaluate chemical insecticides. However, the tests have evolved over the decades and are now designed to evaluate the risks of Bt, specifically the infectivity of the bacteria and toxicological properties of proteins used as active ingredients. The tests are grouped into three tiers, I-III (Betz et al., 1990). Tier I consists of a series of tests aimed primarily at determining whether an isolate of a Bt subspecies, as the unformulated material, poses a risk if used at high levels, typically at least 100 times the amount recommended for field use, to different classes of non-target organisms (Table 2). The principal tests include acute oral, acute pulmonary (inhalation), and acute intraperitoneal evaluations of the material against different vertebrate species, with durations from a week to more than a month, the length depending on the organism. In the most critical tests, the mammals are fed, injected with, and forced to inhale millions of Bt cells in a vegetative or sporulated form. Against invertebrates, the tests are primarily feeding and contact studies. Representative non-target vertebrates and invertebrates include mice, rats, rabbits, guinea pigs, various bird species, fish, predatory and parasitic insects, beneficial insects such as the honeybee, aquatic and marine invertebrates, and plants. If infectivity or toxicity clearly results in any of these tests, then the candidate bacterium would be rejected.

If uncertainty exists after Tier I testing, then Tier II tests must be conducted. These tests are similar to those of Tier I, but require multiple consecutive exposures, especially to organisms where there was evidence of toxicity or infectivity in the Tier I tests, as well as tests to determine if and when the bacterium was cleared from non-target tissues. If infectivity, toxicity, mutagenicity, or teratogenicity is detected, then Tier III tests must be undertaken. These consist of tests such as two-year feeding studies and additional testing of teratogenicity and mutagenicity.

TABLE 2. Tier I Safety Tests Required for Registration of Bt Insecticides in the United States*

Toxicology	Non-Target Organisms/Environmental Fate
Acute oral exposure	Avian oral exposure
Acute dermal exposure	Avian inhalation
Acute pulmonary exposure	Wild mammal
Acute intravenous exposure	Freshwater fish
Primary eye irritation	Freshwater aquatic invertebrates
Hypersensitivity	Estaurine and marine animals
Non-target plants	
Non-target insects including honeybees	

*Adapted from Betz et al. (1990).

The tests can be tailored to further evaluate the hazard based on the organisms in which hazards were detected in the Tier I and II tests.

To date, *none* of the registered bacterial insecticides based on Bt have had to undergo Tier II testing. In other words, no moderate or significant hazards or risks have been detected with any Bt subspecies against any of the non-target organisms studied. As a result, all Bt insecticides are exempted from a food tolerance requirement, i.e., a specific level of insecticide residue allowed on a crop just prior to harvest. Moreover, no washing or other requirements to reduce levels consumed by humans are required. In fact, Bt insecticides can be applied to crops such as lettuce, cabbage, and tomatoes just prior to harvest. It is important to note that such a statement cannot be made for just about any chemical insecticide. This does not mean that registered bacterial insecticides do not have any negative impacts on non-target populations or cannot harm non-target organisms, but rather that these materials do not pose a significant or long-term risk to populations of these organisms. With respect to Bt's safety to humans, which is unparalleled among commonly used insecticides, more detailed treatment of the subject can be found in the article, by J. Kough (pp. 1-10), as well as a recent review by Siegel (2001).

SAFETY OF Bt TO NON-TARGET INVERTEBRATES

The concept of a non-target organism is a relative one and therefore requires some clarification. The term non-target organism generally refers to organisms outside the main target group of pests against which a crop is being protected. For example, with most organophosphate, carbamate, and pyrethroid insecticides, because they often are capable of killing many different types of insects as well as other invertebrates such as spiders and crustacea, non-target organism often refers to non-insect species, but can also be considered to include the many vulnerable insects which do not attack the protected crop. The non-target organisms vulnerable to toxicity from chemical pesticides is often very broad.

With Bt insecticides, owing to their high selectivity, vulnerable non-target organisms are typically much more restricted, including only insects within the taxonomic order to which the primary target insects belong. Bt insecticides are so specific, that their spectrum of activity is generally identified in a very narrow manner, such as "lepidopteran-active," "dipteran-active," or "coleopteran-active." Bt insecticides specificity is often even more limited, such that a Bt subspecies characterized as "lepidopteran-active" may be highly toxic to some lepidopteran species, but have only low or no toxicity to others.

This point can be illustrated with the HD1 isolate of *B. t.* subsp. *kurstaki* (Btk), the isolate used widely in commercial formulations to control lepidopteran pests. Btk is highly toxic and very effective against larvae of the cabbage looper, *Trichoplusia ni*, a common pest of vegetable crops, but typically exhibits poor activity against the beet armyworm, *Spodoptera exigua*, another important caterpillar pest. This is because none of the ICPs produced by Btk (Cry1Aa, Cry1Ab, Cry1Ac, and Cry2A) is very toxic to *Spodoptera* species (Table 1). For this reason, the product XenTari® based on *B. t.* susbp. *aizawai*, which produces an ICP (Cry1Ca) of moderate toxicity to *Spodoptera* species, was developed to control species of this genus.

As noted above, a high degree of specificity and thus safety is attributed to each Bt insecticide, meaning that a Bt subspecies which serves as the active ingredient is limited to being toxic primarily to the insect species of only one taxonomic order. Nevertheless, this would still mean that many non-target species of this order would be sensitive to the Bt endotoxins by the normal route of entry, i.e., ingestion. What we consider a pest is an arbitrary concept as opposed to one based on taxonomy. This has led to considerable misunderstanding about the effects of "lepidopteran-active" Bt subspecies used as insecticides, or the proteins derived from these that are used in Bt crops. An isolate like HD1 of *B. t.* subsp. *kurstaki* has a broad insecticidal range against lepidopteran species, due primarily to the four Cry1 ICPs it produces (Table 1). Therefore, when used in the field it will be capable of killing larvae of target pests as well as certain non-target lepidopterans. Among the targets are larvae of many moth species, especially those of the family Noctuidae (e.g., the corn earworm, the cotton budworm and bollworm, and the cabbage looper). Among the non-targets in certain geographical areas are the larvae of non-pest lepidopterans including larvae of the monarch butterfly, and many other species of moths and butterflies, some of which are endangered species. This can pose a dilemma for farmers as well as the governmental agencies, both regulatory agencies and local governments, in making decisions about the effects of microbial Bt insecticide formulations, and now Bt crops, on non-target organisms.

With respect to specific evaluations of Bt insecticides against non-target invertebrates, there have been numerous studies in the laboratory as well as in field situations under operational pest and vector control conditions. Literally thousands of tons of Bt insecticides have been applied in the environment over the past four decades, and the overall record, especially considering the amounts applied, is one of remarkable safety. The key results of these studies are summarized below.

Microbial insecticide formulations based on different subspecies of Bt have been tested extensively in the laboratory against non-target invertebrates to meet registration requirements, and have also been evaluated in field situations to assess effects of formulated products under operational conditions. Both short-term (i.e., from a few days to several weeks) as well as long-term studies of more than a year have been conducted. In the laboratory studies, doses used to evaluate the effects on non-targets are typically as much as 1,000-fold the amount that these invertebrates would encounter in the field, and in many cases the doses are much higher. Representative non-target invertebrates that have been studied include earthworms, and microcrustacea such as daphnids and copepods that make up much of the zooplankton in treated areas. In addition, insects tested have included non-target Coleoptera (beetles), Diptera (flies), Neuroptera (lacewings), Odonata (dragonflies and damselflies), Trichoptera (caddisflies), and Hymenoptera (parasitic wasps), especially species that constitute the major predator and parasitoid groups that attack the insect pests or disease vectors that are the targets of the Bt applications. Larvae and adults of beneficial insects such as the honeybee, *Apis mellifera*, are also tested. In testing Bt products used against caterpillar pests, more emphasis has been placed on evaluating the effects on terrestrial non-target invertebrates. However, because these products can drift or be washed into streams and ponds, many aquatic invertebrates have been tested in laboratory studies and in natural habitats. In the case of *B. t.* subsp. *israelensis*, used to control mosquito and blackfly larvae, greater emphasis has been placed on evaluating the effects on aquatic non-target insects and other arthropods.

Summaries of these results and those of other studies carried out over the past thirty years show virtually no adverse direct or indirect effects, especially long term effects, of Bt or formulated products of Bt on non-target populations. The obvious exceptions are non-target species that are closely related to the target pests or vectors, or insects such as endoparasitic hymenopteran species that require the target lepidopteran pests as hosts. But even these are not affected in some cases. Moreover, even in "forced" feeding studies, Bt subspecies did not have an effect on insects or non-target invertebrates, such as shrimp, that were outside the order of insects designated as the target group (see Glare & O'Callaghan, 2000, for a comprehensive summary of these studies). In some of the earliest studies, effects were seen on earthworms and flies. But these early studies were conducted with strains that may have had the β-exotoxin, which has not been permitted in commercial formulations for decades.

In cases where the effects of Bt on non-target populations have been monitored under field conditions, these were much less than those resulting from the use of chemical insecticides. The effects of Bt therefore must be viewed from the perspective of the consequences of using alternative control technologies, unless one considers regular devastation of food crops and vectoring of diseases by insect pests a permissible alternative. An appropriate example is the use of *B. t.* subsp. *israelensis* (Bti) in the Volta River Basin to control the larvae of *Simulium damnosum*, the blackfly vector of Onchocerciasis, a blinding eye disease of humans. The Onchocerciasis control program is sponsored by the World Health Organization and United Nations Development Program. After more than a decade of intensive use, it was concluded that Bti was of "only the slightest of hazards" to any of the non-target organisms tested. More specifically, when Bti formulations were applied to rivers, the "drift" of invertebrates, i.e., the target and non-target invertebrates found floating in the rivers and presumably killed or disturbed by the application, increased 2- to 3-fold in comparison to untreated rivers. However, when chemical insecticides were applied under similar ecological conditions, the drift increased 20- to 40-fold. In other words, the application of chemical insecticides was approximately ten times more detrimental to the non-target invertebrate populations than the use of Bti. In addition to the much greater impact of the chemical insecticides on non-target invertebrates in the rivers, the blackfly population began to develop resistance to these chemicals. Replacement of the latter with Bti-based insecticides during the drier periods of the year ensured the success of this program, and allowed large fertile areas of the river valleys in West Africa to be returned to productive agriculture with the renewal of irrigation that once had threatened public health. No resistance has thus far developed to Bti-based insecticides.

SAFETY OF Bt CROPS
TO NON-TARGET INVERTEBRATES

Present Bt crops represent an early phase of a new technology, and it is easy to exaggerate their potential benefits and shortcomings. Given the forty-year safety record of Bt insecticides along with the well-accepted empirical methods for testing the safety of chemical and bacterial insecticides, it is appropriate that a combination of prior studies and empirical methods be used to establish the safety or lack thereof of Bt crops. Over the past few years, studies of Bt crop safety using empirical methods have begun to appear in the scientific literature. These studies have examined the effects on non-target invertebrates and vertebrates including mammals in the laboratory and field. Under operational growing conditions, these studies show that Bt crops have no significant adverse consequences for non-target invertebrate populations, and if anything their use is beneficial because the use of broad spectrum chemical insecticides is reduced (Betz et al., 2000; Carpenter et al., 2002). Replacement of chemical pesticides with Bt crops provides better protection of beneficial insect populations due to the much greater specificity of Bt proteins. Furthermore, a measure of safety is provided, even for non-target insects that might be susceptible to the Bt component(s) of a transgenic crop, in that exposure to toxicity requires that the organism eat crop tissues.

Cry proteins produced by transgenic plants (Table 3) are not easily extractable in the amounts that would be required for studies designed to test the effects on non-target organisms. Thus, the effects are currently assessed by feeding test species Cry proteins produced in either

TABLE 3. Cry Proteins Produced by Bt Crops Registered in the United States

Crop[*,†]	Cry Protein[‡]	Target Insects
Cotton	Cry1Ac	Tobacco budworm, *Heliothis virescens* Cotton bollworm, *Helicoverpa zea* Pink bollworm, *Pectinphora gossypiella*
Corn	Cry1Ab	European corn borer, *Ostrinia nubilalis* Southwestern corn borer, *Diatraea grandiosella* Corn earworm, *Helicoverpa zea*
Corn	Cry1Ac	European corn borer, *Ostrinia nubilalis* Southwestern corn borer, *Diatraea grandiosella*
Potato	Cry3Aa	Colorado potato beetle, *Leptinotarsa decemlineata*

[*]Source: U.S. Environmental Protection Agency (*http://www.epa.gov/oppbppdl/biopesticides/factsheets/*).
[†]See Schuler et al. (1998) for other plants engineered to produce Bt Cry proteins.
[‡]Most of these proteins are produced as full-length molecules similar in mass to those produced by the Bt subspecies from which they were derived. Some Bt corn varieties based on Cry1Ab produce a truncated version of this protein.

Escherichia coli, a *Bacillus* species, or on various Bt-crop tissues such as leaves or pollen. Tests of Cry proteins produced in *E. coli* are similar to those used to evaluate these proteins when produced by *B. thuringiensis*, except that in many cases an activated form of the toxin is used to produce what could be considered a "worst case" hazard assessment. To complement laboratory studies, several field studies have been conducted in which non-target insect populations were monitored on Bt crops, mainly Bt maize and Bt cotton, throughout the growing season.

Most of the laboratory studies have been performed in the United States, where a complex of non-target organisms serves as a standard group for which results are accepted by the U.S. Environmental Protection Agency. The standard test invertebrates have included a range of terrestrial and freshwater aquatic organisms generally considered beneficial. These typically are larval and/or adults of one or more of the following organisms: the honeybee, parasitic wasps, predatory ladybird beetles and lacewings, "soil-dwelling" springtails, earthworms, and as a representative of a freshwater aquatic crustacean, a daphnid. In these tests, the non-target organisms were typically exposed to or fed amounts of toxin that were in the range of at least a hundred to several thousand times the amount they would be exposed to or consume under natural conditions. In such tests, when no effects are observed at the highest dose or rate tested, this amount is referred to as the no-observed-effect-level, NOEL. For a crop like Bt corn, the amount of Cry protein in a maturing field is estimated to be about 500 g per hectare, and thus the test levels are adjusted to ensure a dose at least 1,000 times this level. To date, no significant effects have been found on non-target invertebrates and vertebrates evaluated in these studies. As and example, the results of several tests carried out with the Cry1Ab protein are summarized in Table 4. Most of the studies conducted so far have been short-term studies, lasting from several days to a few weeks, though several multi-year studies are now underway. The results of the first studies carried out in the 1990s, summarized in Table 4, have shown no adverse effects of Bt Cry proteins on the organisms tested.

These initial studies indicated that Bt crops would be safe for most non-target invertebrates. However, several reports of detrimental effect of Bt Cry proteins and Bt crops questioned the putative safety of Bt crops. The most widely publicized of these studies was a study of the potential impact of Bt corn on larvae of the monarch butterfly, *Danaus plexippus*, showing that these are sensitive to Bt corn (Cry1Ab) pollen (Losey et al., 1999). In this laboratory study, milkweed leaves were covered with Bt corn pollen and then fed to larvae. Control larvae were

TABLE 4. Cry1Ab Effects on Non-Target Invertebrates and Non-Mammalian Vertebrates

Non-Target Organism	No Effects Level[*,†]
Invertebrates	
Insects	
Honeybee, *Apis mellifera*	20 ppm
Ladybird beetles, *Hippodamia convergens*	20 ppm
Green Lacewing, *Chrysoperla carnea*	16 ppm
Wasp parasite, *Brachymeria intermedia*	20 ppm
Springtail, *Folsomia candida*	50 µg/g leaf tissue
Earthworms	
Earthworm, *Eisenia fetida*	200 mg/kg soil
Freshwater crustacea	
Daphnid, *Daphnia magna*	100 mg pollen/liter
Vertebrates[‡]	
Northern bobwhite quail	100,000 ppm
Channel catfish	< 3 µg/g maize feed
Broiler chickens	< 3 µg/g maize feed

[*]Adapted from Sanders et al. (1998), Brake and Vlachos (1998), Yu et al. (1997), and "Factsheets" produced by the U.S. Environmental Protection Agency and available on the following Agency website: (*http://www.epa.gov/oppbppdl/biopesticides/factsheets/*). Results of similar tests on other Cry proteins used in Bt crops are similar, and can be viewed on the above website.

[†]Tests are conducted using a single high level of toxin much higher than that estimated the test organisms would likely encounter under field conditions. This is referred to as the no-observed-effect-level (NOEL).

[‡]Fed Bt corn grain.

fed on milkweed leaves covered with non-Bt pollen or untreated milkweed. The key finding of the study was that the larvae fed milkweed leaves treated with unquantified amounts of Bt pollen had a lower survival rate (56%) in comparison to the controls (100%). The authors reported that their results had "potentially profound implications for the conservations of monarch butterflies" because the central corn belt, where Bt corn adoption by farmers continues to grow, is also an important habitat for monarchs that migrate to the U.S. from Mexico each year.

In assessing the relevance of the findings on monarch larvae, or other non-target organisms for that matter, it should be kept in mind that bacterial insecticides based on Bt should be just as toxic if not more so. This is because Bt formulations are more complex than Bt crops, containing multiple Cry proteins and synergists as well as spores and formulation ingredients. In other words, monarch larvae that feed under field conditions on milkweed leaves treated with a product that contains *B. thuringiensis* subsp. *kurstaki*, from which the Cry1Ab protein gene was derived, will be equally if not more sensitive to the microbaial insecticide formulation. Similarly, predators that feed on caterpillars intoxicated as a result of feeding on a Bt formulation will be no less sensitive to the activated toxins in these larvae. So the issue may be viewed less as one of Bt crops, but more whether Cry proteins will impact beneficial insects regardless of the source.

Extraordinary attention was given to the preliminary findings in the scientific and popular press on the potential negative effects of Bt pollen on monarch populations. A benefit of this attention was that it resulted in a series of collaborative studies in 1999 and 2000 devoted to a much more rigorous assessment of these potential negative effects under field conditions throughout the U.S. corn belt and Canada (Hellmich et al., 2001; Sears et al., 2001). The overall conclusion of these studies was that the effects of Bt corn on monarch populations will be negligible, especially in comparison to the effects of using chemical insecticides to control corn pests. In part this is a result of low Cry protein levels that occur in most currently marketed varieties of Bt corn (Sears et al., 2001). However, even in cases where high levels of Cry1Ab are produced in pollen, the overall impact on monarch populations would likely be negligible. This is because pollen is only shed during a limited period of the corn-growing season, use of Bt corn reduces the use of chemical insecticides, and milkweed, the host plant of monarch larvae, grows in many areas where Bt corn is not grown. In a similar recent study carried out under field conditions, it was also found that Bt corn pollen would not likely have any significant impact on populations of the black swallowtail, *Papillio polyxenes* (Wraight et al., 2000).

In other less popularized studies of the effects of Bt proteins on non-target invertebrates, it was shown that immature lacewings (*C. carnea*) fed on prey that had eaten Bt maize (Cry1Ab) suffered greater mortality than control lacewings fed on prey that had eaten non-Bt maize (Hilbeck et al., 1998a). Only 37% of the lacewings survived when fed larvae of the cotton leafroller, *Spodoptera littoralis*, or the European corn borer, *Ostrinia nubilalis* that had eaten Bt maize. By comparison 62% of the control group survived when fed on caterpillars that had eaten non-Bt maize. In a subsequent study, using an artificial liquid diet it was determined that immature *C. carnea* were sensitive to the Cry1Ab toxin at a level of 100 µg per ml of diet (Hilbeck et al., 1998b).

However, the level of Cry1Ab in maize is about 4 µg/g fresh weight, which is considerably less than 100 µg/ml (Hilbeck et al., 1998b). At present there appears to be no explanation for the discrepancy between the artificial diet and exposure via prey. More recent laboratory studies have confirmed that *C. carnea* fed *S. littoralis* larvae fed on Bt corn suffered higher mortality and longer developmental times than *C. carnea* control larvae (Dutton et al., 2002).

Regardless of the whether the results obtained against non-target organisms in laboratory studies show favorable, unfavorable, or neutral effects, it will be long term studies under field conditions that reveal effects with any relevance, if such exist. The reason is that laboratory studies are designed to reveal any potential adverse effects by exposing non-target organisms to excessively high levels of Bt proteins, levels that would not be encountered under field conditions. Moreover, field studies should include comparisons to current agricultural practices, which often will include the use of chemical insecticides. At present, there have only been a few studies that have evaluated the effects of Bt crops on non-target organisms under field conditions over the length of the growing season. These studies are nevertheless important because non-target organisms and their prey were exposed to Bt Cry proteins in the form synthesized in the crop and over a continuous period at an operational level. The non-target organisms studied under field conditions have all been insects or spiders and the test crops either Bt corn or Bt cotton. The insects consisted of a plant pest and four beneficial insects, specifically two parasites and two predators, one of which was the lacewing, *C. carnea*. In these season-long studies, no adverse effects were observed on any of the non-targets under field conditions (Table 5), even on *C. carnea*, which is known to be sensitive to Cry1Ab.

As part of a much more comprehensive program to evaluate the effects of Bt crops on non-target organisms, several large-scale multiyear studies are currently under way in the U.S. using Bt corn and Bt cotton as model systems. Results are not yet published in the peer-reviewed scientific literature, as data are only available for the first two years. However, preliminary reports have indicated that neither of these crops have any significant impact on non-target invertebrates (Moar et al., 2002; Dively & Rose, personal communication). In the Bt cotton studies, carried out in large scale (multi-acre plots) in several states throughout the southeastern United States, no significant differences were found between Bt cotton and conventional cotton not treated with insecticides. Similar results have been obtained in Bt cotton plantings in the Arizona (Naranjo, personal communication). In the studies of Bt

TABLE 5. Effects of Bt Crops on Non-Target Invertebrates Under Field Conditions

Non-Target (Insect Order)	Crop	Cry Protein	Adverse Effects	Reference
Lygus lineolaris (Heteroptera)	Cotton	Cry1Ac	None	Hardee & Bryan (1997)
Coleomegilla maculata (Coleoptera)	Corn	Cry1Ab	None	Pilcher et al. (1997)
Orius insidiosus (Heteroptera)	Corn	Cry1Ab	None	Pilcher et al. (1997)
Chrysoperla carnea (Neuroptera)	Corn	Cry1Ab	None	Pilcher et al. (1997)
Eriborus tenebrans (Hymenoptera)	Corn	Cry1Ab	None	Orr & Landis (1997)
Macrocentrus grandi (Hymenoptera)	Corn	Cry1Ab	None	Orr & Landis (1997)

corn, in which fresh corn is used as the model crop in mid-Atlantic states, again no significant detrimental impacts of Bt fresh corn have been observed on non-target invertebrate populations over a period of two years. In the latter studies, over 100 species of non-target organisms are being monitored, including predaceous ground beetles, collembola, aphids, parasitic wasps, chrysopids, and many species of spiders. If these initial results are confirmed though ongoing studies, Bt crops will be shown to represent a significant improvement in pest control technology, one that is safe for humans as well as other non-target organisms.

CONCLUSIONS

Based on a wide range of scientific evidence, Bt crops are a novel and safe pest control technology that will improve agricultural ecosystems because their spectrum of activity against insects and other non-target invertebrates is so much narrower than synthetic chemical insecticides. To date, other than obligate parasites such as parasitic wasps dependent on the target pests as hosts, no detrimental impacts have been found on non-target insect populations or other invertebrates under operational growing conditions in Bt cotton or Bt corn fields. Preliminary studies have identified some negative effects of Bt crops on non-target insects under laboratory conditions, but subsequent field studies have shown that the risk to populations of these under field conditions is negligible. In addition, laboratory studies conducted against a range of non-target vertebrates have shown no detrimental effects from eating Bt crops.

Although the public remains concerned about the safety of Bt crops owing to negative reports about these in the popular press, there is no evidence that Bt crops pose risks for humans any greater than those that result from eating non-genetically engineered crops. Thus, as former U.S. President Jimmy Carter (1998) wrote several years ago, the panic over genetically modified plants is completely uncalled for. Similar views have been expressed by many scientific organizations with expertise about Bt crops, as well as by the U.S. National Academy of Sciences (2002). While there is always the possibility that these crops will have some detrimental effects, these remain to be identified. Risk-benefit analyses show that the benefits of using Bt crops far outweigh the risks. Further deployment of this new technology should reduce significantly the use of synthetic chemical insecticides, to name just one substantial benefit. Despite the safety of Bt crops, other scientific, economic and political controversies over the use of this technology will delay its implementation in many countries for years to come. Ultimately, however, transgenic crops are here to stay, and the use of this technology will expand worldwide. The technology is simply too powerful and offers to many benefits to humanity not to be developed further and used.

REFERENCES

Betz, F. S., S. F. Forsyth and W. E. Stewart. (1990). Registration requirements and safety considerations for microbial pest control agents in North America, pp. 3-10. In "Safety of Microbial Insecticides." M. Laird, L. A. Lacey, L. A., and E. W. Davidson, Editors. CRC Press, Inc., Boca Raton.

Betz, F. S., B. G. Hammond and R. L. Fuchs. (2000). Safety and advantages of *Bacillus thuringiensis*-protected plants to control insect pests. *Regulatory Toxicology and Pharmacology* 32:156-173.

Brake, J. and D. Viachos. (1998). Evaluation of transgenic event 176 Bt-corn in broiler chickens. *Poultry Science* 77: 648-653.

Bravo, A., J. Sanchez, T. Kouskoura and N. Crickmore. (2002). N-terminal activation is an essential early step in the mechanism of action of the *Bacillus thuringiensis* Cry1Ac insecticidal toxin. *Journal of Biological Chemistry* 27: 23985-23987.

Carpenter, J., A. Felsot, T. Goode, M. Hamming, D. Onstad and S. Sankula. (2002). Comparative environmental impacts of biotechnology-derived and traditional soybean, corn, and cotton crops. Council for Agricultural Science and Technology, Ames, Iowa. 189 pp. (www.cast-science.org).

Carter, J. (1998). Panic over genetically modified plants completely uncalled for. *New York Times*, August 26.

Dutton, A., H. Klein, J. Romeris and F. Bigler. (2002). Uptake of Bt-toxin by herbivores feeding on transgenic maize and consequences for the predator *Chrysoperla carnea*. *Ecological Entomology* 27: 441-447.

Federici, B. A. (1999). *Bacillus thuringiensis* in Biological Control. Chapter 21, pp. 519-529. In "Handbook of Biological Control." T. S. Bellows, G. Gordh, and T. W. Fisher, Editors. Academic Press, Inc., San Diego.

Glare, T. R. and M. O'Callaghan. (2000). *Bacillus thuringiensis*: Biology, Ecology and Safety. John Wiley & Sons, LTD, Chichester.

Hardee, D. D. and W. W. Bryan. (1997). Influence of *Bacillus thuringiensis*-transgenic and nectariless cotton on insect populations with emphasis on the tarnished plant bug (Heteroptera: Miridae). *Journal of Economic Entomology* 90: 663-668.

Hellmich, R. L., B. D. Siegfried, M. K. Sears, D. E. Stanley-Horn, M. J. Daniels, H. R. Mattila, T. Spencer, K. G. Bidne and L. Lewis. (2001). Monarch larvae sensitivity to *Bacillus thuringiensis*-purified proteins and pollen. *Proceedings of the National Academy of Sciences U.S.A.* 98: 11925-11930.

Hilbeck, A., M. Baumgartner, P. M. Fried and F. Bigler. (1998a). Effects of transgenic *Bacillus thuringiensis* corn-fed prey on mortality and development time of immature *Chrysoperla carnea* (Neuroptera: Chrysopidae). *Environmental Entomolology* 27: 480-487.

Hilbeck, A., W. J. Moar, M. Pustai-Carey, A. Filippini and F. Bigler. (1998b). Toxicity of *Bacillus thuringiensis* Cry1A(b) toxin to the predator *Chrysoperla carnea* (Neuroptera: Chrysopidae). *Environmental Entomolology* 27: 1255-1263.

Hofte, H. and H. R. Whiteley. (1989). Insecticidal crystal proteins of *Bacillus thuringiensis*. Microbiological Reviews 53(2):242-255.

Knowles, B. H. and J. A. T. Dow. (1993). The crystal δ-endotoxins of *Bacillus thuringiensis*: Models for their mechanism of action on the insect gut. *BioEssays* 15: 469-476.

Li, J., J. Carroll and D. J. Ellar. (1991). Crystal structure of insecticidal δ-endotoxin from *Bacillus thuringiensis* at 2.5 angstrom resolution. *Nature* 353: 815-821.

Losey, J. J., L. Raynor and M. E. Cater. (1999). Transgenic pollen harms monarch larvae. *Nature* 399: 214.

Moar, W. J., M. Eubanks, B. Freeman, S. Turnipseed, L. Ruberson and G. Head. (2002). Effects of Bt cotton on biological control agents in the southeastern United States. First International Symposium on Biological Control. Honolulu, Hawaii. R. Van Driesche, Editor. U.S. Department of Agriculture, Forest Service, Morgantown, West Virginia.

Noteborn, H. P. J. M., M. E. Bienenmann-Ploum, J. H. J. van den Berg, G. M. Alink, L. Zolla, A. Reynaerts, M. Pensa and H. A. Kuiper. (1995). Safety Assessment of *Bacillus thuringiensis* insecticidal protein CRYIA(b) expressed in transgenic tomato, pp. 134-147. In K.-H. Engel, G. R. Takeoka and R. Teranishi (eds.), Genetically Modified Foods: Safety Issues. American Chemical Society, Washington, DC.

Orr, D. B. and D. A. Landis. (1997). Oviposition of European corn borer (Lepidoptera: Pyralidae) and impact of natural enemy populations in transgenic versus isogenic corn. *Journal of Economic Entomology* 90: 905-909.

Pilcher, C. D., J. J. Obrycki, M. E. Rice and L. C. Lewis. (1997). Preimaginal development, survival, and field abundance of insect predators on transgenic *Bacillus thuringiensis* corn. *Environonmental Entomology* 26: 446-454.

Sanders, P. R., T. C. Lee, M. E. Groth, J. D. Astwood and R. L. Fuchs. (1998). Safety assessment of insect-protected corn, pp. 241-256. In Thomas, J. A. (ed.), Biotechnology and Safety Assessment, 2nd ed. Taylor & Francis, Ltd, London.

Schnepf, E., N. Crickmore, J. Van Rie, D. Lereclus, J. Baum, J. Feitelson, D. R. Zeigler and D. H. Dean. (1998). *Bacillus thuringiensis* and its pesticidal crystal proteins. *Microbiology and Molecular Biology Reviews* 62: 775-806.

Schuler, T. H., G. M. Poppy, B. R. Kerry and I. Denholm. (1998). Insect-resistant transgenic plants. *Trends in Biotechnology* 16: 168-175.

Sears, M. K., R. L. Hellmich, D. E. Stanley-Horn, K. S. Oberhauser, J. M. Pleasants, H. R. Mattila, S. D. Siegfried and G. P. Dively. (2001). Impact of Bt corn pollen on monarch butterfly populations: a risk assessment. *Proceedings of the National Academy of Sciences U.S.A.* 98: 11937-11942.

Siegel, J. P. (2001). The mammalian safety of *Bacillus thuringiensis*-based insecticides. *Journal of Invertebrate Pathology* 77: 13-21.

U.S. National Academy of Sciences: Gould, F. L., D. A. Andow, B. Blossey, I. Chapela, N. C. Ellstrand, N. Jordan, K. R. Lamkey, B. A. Larkeins, D. K. Letourneau, A. McHughen, R. L. Philips and P. B. Thrompson. (2002). Environmental Effects of Transgenic Plants. National Academy Press. Washington, DC.

Wraight, C. L., A. R. Zangerl, M. J. Carroll and M. R. Berenbaum. (2000). Absence of toxicity of *Bacillus thuringiensis* pollen to black swallowtails under field conditions. *Proceedings of the National Academy of Sciences U.S.A.* 97: 7700-7703.

Yu, L., R. R. Berry and B. A. Croft. (1997). Effects of *Bacillus thuringiensis* toxins in transgenic cotton and potato on *Folsomia candida* (Collembola: Isotomidae) and *Oppia nitens* (Acari: Orbatidae). *Ecotoxicology* 90:113-118.

The Role of Microbial Bt Products in U.S. Crop Protection

Kathleen Walker
Michael Mendelsohn
Sharlene Matten
Marvin Alphin
Dirk Ave

SUMMARY. Microbial *Bacillus thuringiensis* (Bt) insecticides have been used for over 40 years. In the United States, Bt formulations are primarily applied to control lepidopteran pests on fruit and vegetable crops, to control gypsy moth in forests and to control dipteran pests (mosquitoes and blackflies) that bite humans. A highly selective insecticide with activity conferred primarily by insecticidal crystal proteins (ICPs), Bt is generally not harmful to humans, non-target wildlife or beneficial ar-

Kathleen Walker is affiliated with the Department of Entomology, 410 Forbes, College of Agriculture and Life Sciences, University of Arizona, Tucson, AZ 85721-0036 (E-mail: krwalker@ag.arizona.edu).

Michael Mendelsohn and Sharlene Matten are affiliated with the U.S. Environmental Protection Agency, OPP-BPPD, 1200 Pennsylvania Avenue, NW (7511C), Washington, DC 20460.

Marvin Alphin and Dirk Ave are affiliated with Valent BioSciences Corporation, 870 Technology Way, Suite 100, Libertyville, IL 60048.

The views expressed in this article are those of the authors and do not necessarily represent those of the United States Environmental Protection Agency (EPA). The use of trade, firm or corporation names in this article is for the information and convenience of the reader. Such use does not constitute an official endorsement or approval by EPA of any product or service to the exclusion of others that may be suitable.

[Haworth co-indexing entry note]: "The Role of Microbial Bt Products in U.S. Crop Protection." Walker, Kathleen et al. Co-published simultaneously in *Journal of New Seeds* (Food Products Press, an imprint of The Haworth Press, Inc.) Vol. 5, No. 1, 2003, pp. 31-51; and: *Bacillus thuringiensis: A Cornerstone of Modern Agriculture* (ed: Matthew Metz) Food Products Press, an imprint of The Haworth Press, Inc., 2003, pp. 31-51. Single or multiple copies of this article are available for a fee from The Haworth Document Delivery Service [1-800-HAWORTH, 9:00 a.m. - 5:00 p.m. (EST). E-mail address: docdelivery@haworthpress.com].

http://www.haworthpress.com/store/product.asp?sku=J153
© 2003 by Taylor & Francis.
10.1300/J153v05n01_03

thropods. Its selectivity and unique mode of action make it an important alternative to conventional chemical insecticides, and many integrated pest management (IPM) programs for particular fruit and vegetable crops as well as certified organic production include the use of Bt. Agricultural commercialization and adoption of plant-incorporated Bt presents new opportunities to expand the use of Bt ICPs for agricultural pest control, but also raises concerns about the potential for accelerated development of pest resistance to Bt. The relative risks and benefits of microbial and plant-incorporated Bt products are introduced. *[Article copies available for a fee from The Haworth Document Delivery Service: 1-800-HAWORTH. E-mail address: <docdelivery@haworthpress.com> Website: <http://www.HaworthPress.com> © 2003 by The Haworth Press, Inc. All rights reserved.]*

KEYWORDS. *Bacillus thuringiensis* (Bt), microbial formulation, integrated pest management (IPM), insecticidal crystal protein (ICP)

INTRODUCTION

Discovered in the early 20th century, the bacterium *Bacillus thuringiensis* (Bt) is the most widely used microbially-derived insecticide today. Microbial Bt products consisting of bacterial spores and associated protein crystals have been applied to crops for over 40 years. Due to the complex mode of action of its toxins, Bt is generally safe for vertebrates and many beneficial insects (Flexner et al. 1986) and is considered more resilient to the development of insect resistance than most chemical insecticides that rely on neurotoxicity (Lüthy et al. 1982; Croft 1990). The primary insect killing activity is conferred by the Bt insecticidal crystal proteins (ICPs). These qualities make microbial Bt a valuable tool for integrated pest management (IPM), particularly in fruit and vegetable crops, and for organic production. Based on the advice of the 1996 Pesticide Program Dialogue Committee, an advisory committee to the U.S. Environmental Protection Agency (EPA), the EPA determined that pest susceptibility to Bt is in the "public good" and should be protected against efficacy loss due to the development of resistant insects (U.S. EPA 1996, 1998a, 2001). The recent commercial development and wide-spread adoption of transgenic corn and cotton crops containing Bt ICP genes raise questions about the future of microbial Bt. In this paper we discuss the utility and current patterns of use of microbial Bt products in the United States and introduce the relative benefits and risks of microbial vs. transgenic Bt in different cropping systems.

BACKGROUND

Bacillus thuringiensis occurs naturally throughout the world and has been found in dead insects, insect breeding environments, stored grain, soil and leaf surfaces (Dulmage & Aizawa 1982). Although Bt is usually found in association with insects, epizootic infections are rare (Aronson & Shai 2001). A decade after the initial discovery of Bt in diseased silkworm larvae in Japan by S. Ishiwata, the bacterium was isolated from diseased flour moth larvae in Thuringia, Germany by E. Berliner (1915). Later, E. Kurstak of France and H. Dulmage of the United States Department of Agriculture isolated Bt strains active against Lepidoptera that were grouped together and given the subspecies name *kurstaki* (Dulmage & Aizawa 1982). Goldberg and Margalit (1977) discovered *Bacillus thuringiensis* subspecies *israelensis* in Israel within soil samples taken from mosquito breeding sites. Not surprisingly, subspecies *israelensis* is effective against mosquitoes and black flies. Krieg et al. (1983) discovered *Bacillus thuringiensis* subspecies *tenebrionis* in Germany. This subspecies was isolated from diseased yellow mealworm (*Tenebrio molitor*) larvae and is effective against certain Coleopteran insects pests. Numerous strains of Bt have been isolated, and many more may yet be discovered (Dent 1993). Other strains have been developed in the laboratory via bacterial mating, transconjugation, and the techniques of molecular biology.

Bt produces a number of toxins that contribute towards its efficacy against target pests. These include the Crystal (Cry) and Cytolytic (Cyt) ICPs, also known as δ-endotoxins, heat stable α-exotoxins, heat labile α-exotoxins, and vegetative insecticidal proteins. The primary insecticidal toxins in Bt are the Cry (Lüthy et al. 1982) ICPs. ICPs are present as protoxins in crystal inclusions on the outside of the bacterial spore. The insect must first ingest the protoxins, which are then solubilized in the insect's digestive tract. The lepidopteran and dipteran active protoxins require an alkaline insect gut for solubillization. The coleopteran protoxin Cry3A solubilizes in a neutral or slightly acidic insect gut, and this may be accounted for by a lack of disulphide bonds in the protein structure (Knowles 1994). Solubilized ICPs activated by proteolytic digestion become bound to the midgut membrane causing the formation of a pore in the membrane. Following pore formation lysis of the gut contents occurs, leading to sepsis and finally death of the insect (Aronson & Shai 2001).

DEVELOPMENT AND CURRENT USE
OF MICROBIAL Bt INSECTICIDES

A range of microbial Bt products have been developed since the initial discoveries. Bt was first available commercially in France in 1938 and in the United States during the 1950s (Lüthy et al. 1982). The early strains of Bt used in the United States contained *Bacillus thuringiensis* subspecies *thuringiensis* and were low in potency. The *kurstaki* subspecies group became available commercially in the United States in the 1970s. Dulmage's strain, HD-1, made it possible to compete against chemical insecticides because of its efficacy and price (Dulmage 1970). Currently, there are 13 active ingredients in 123 microbial pesticide products registered for use in the United States by the U.S. Environmental Protection Agency (Table 1). The vast majority of these are the *kurstaki* subspecies, registered for use against lepidopteran pests in agriculture, forestry, stored products and home and garden. Another significant number of registrations are for the *israelensis* subspecies for use against mosquitoes, black flies, and other dipteran pests. *Aizawai* strains are registered for lepidopteran pests and *tenebrionis* is registered for coleopterans such as the Colorado potato beetle. In 1998 the EPA decided to stop grouping new isolates under the subspecies name because it is now known that the ICP genes, which generally reside on transferable genetic elements (plasmids) can be readily moved from one isolate to another, regardless of subspecies (U.S. EPA 1998b). The EPA may break some of the larger Bt active ingredients currently designated by subspecies only into strain specific active ingredients, but at present some currently listed active ingredients are specific isolates and others are broad subspecies (see Table 1). In the United States today, most of the different Bt subspecies and specific isolates are applied against lepidopteran pests in agriculture and forestry or against dipteran pests (mosquitoes and black flies) to protect public health and reduce nuisance biting.

The narrow spectrum of toxicity and short environmental persistence of microbial Bt insecticides make them safer than many conventional chemical insecticides, but more difficult to use. Bt formulations are applied as fairly inert crystalline proteins that only become activated in the

TABLE 1. Bt Active Ingredients Currently Registered at EPA

Active Ingredient	Number of U.S. products registered	EPA PC code	Target pest order
Bacillus thuringiensis subspecies *kurstaki*	63	006402	Lepidopteran
Bacillus thuringiensis subspecies *israelensis*	23	006401	Dipteran
Bacillus thuringiensis subspecies *tenebrionis*	3	006405	Coleopteran
Bacillus thuringiensis subspecies *aizawai*	5	006403	Lepidopteran
Bacillus thuringiensis subspecies *aizawai* strain GC-91	2	006426	Lepidopteran
Bacillus thuringiensis subspecies *israelensis* strain EG2215	1	006476	Dipteran
Bacillus thuringiensis subspecies *kurstaki* strain EG2348	6	006424	Lepidopteran
Bacillus thuringiensis subspecies *kurstaki* strain EG2371	2	006423	Lepidopteran
Bacillus thuringiensis subspecies *kurstaki* strain EG7841	2	006453	Lepidopteran
Bacillus thuringiensis subspecies *kurstaki* strain EG7826	5	006459	Lepidopteran
Bacillus thuringiensis subspecies *kurstaki* strain DG7673	2	006448 006447	Coleopteran
Bacillus thuringiensis subspecies *kurstaki* strain BMP123	7	006407	Lepidopteran
Bacillus thuringiensis subspecies *kurstaki* strain M-200	2	006452	Lepidopteran

Source: U.S. EPA Office of Pesticide Programs. Updated counts and electronic product labels can be found online at *http://www.cdpr.ca.gov/docs/epa/epamenu.htm*.

midgut of the susceptible insect pest. As such, they do not present a significant health risk to humans or vertebrate wildlife (U.S. EPA 1998c; 1998d) (see J. Kough, this volume, pp. 1-10). Microbial Bt formulations are also considered substantially less toxic than conventional pesticides to important beneficial arthropods including many predators and parasitoids that prey on potential pest species (Flexner et al. 1986) (see B. Federici, this volume, pp. 11-30), although large-scale applications in forestry may impact weed biocontrol agents (James et al. 1993). Many synthetic chemical insecticides kill beneficial arthropods, requiring growers to apply additional insecticides to control secondary pests enabled to emerge after the initial pesticide application (DeBach and Rosen 1990). Another advantage is the lack of cross-resistance between Bt ICPs and chemical insecticides. Bt offers an alternative when pests develop resistance to common organophosphate or carbamate insecticides, which have very similar modes of action (Ware 1989).

Some features that contribute to the efficacy and safety of Bt also make its use more complicated and expensive than many conventional pesticides. Bt ICPs and spores are rapidly inactivated when exposed to UV light. This lack of persistence in addition to its specificity reduces risks of exposure for susceptible non-target insects and reduces the selection for Bt-resistant target pests. The short residual activity of the sprays means correct timing of applications is critical and growers may have to reapply Bt frequently for extended protection. Bt must be eaten *by the target pest* to be effective and must be applied when larvae are actively feeding, preferably during warm, dry weather (UCIPM 1999). Incorrect spray methods or bad weather can compromise efficacy. Finally, Bt sprays cannot reach pests that feed inside plant tissue or below the soil level. Together, these properties make Bt an important tool in integrated pest management, but may limit its cost-effectiveness in some cropping situations.

Use in Agriculture

Table 2 summarizes the use of microbial Bt products on specific agricultural crops in 2000-1 in the United States. The primary crops treated routinely with microbial Bt are fruits and vegetables, commonly referred to as "specialty crops." As field crops, particularly the large commodities such as corn, wheat, cotton and soybeans, dominate agro-chemical consumption and sales in the United States, microbial Bt formulations represent only a small portion of total insecticide sales (between 1 and 2% in 1999) (Nester et al. 2002). Prior to the adoption of plant-incorporated or transgenic Bt cotton in the late 1990s, however, upland cotton was also a significant portion of the microbial Bt market, with about 9% of the total acreage treated with Bt in 1995 (USDA-NASS 2002c). Since that time the percentage of cotton acreage treated with microbial Bt has declined to about 1% in 2000 (USDA-NASS 2002c).

In spite of its minor use in commodity crops, Bt plays a special role in pest management on many fruit and vegetable crops. For example, Bt is

TABLE 2. Summary of Microbial Bt Use on U.S. Agricultural Crops: 2000-1[a]

Crops[b]	Acres planted[c]	Percentage acres treated with Bt	Applications per treated acre	Total value of utilized crop ($1000)	Value of crop treated with Bt ($1000)
almonds	480,000	18	1.6	687,742	123,794
apples	431,200	13	1.5	1,514,301	196,859
apricots	19,430	19	1.4	26,472	5,030
artichokes	8,800	6	3.7	61,021	3,661
beans (snap)	98,700	20	3.1	250,794	50,159
blackberries	6,160	12	1.5	14,042	1,685
blueberries	40,580	8	1.4	165,238	13,219
broccoli	144,500	10	1.5	633,904	63,390
cabbage– fresh processing	82,410 7,740	60 9	4.0 1.3	326,198 9,862	195,719 887
cantaloupes	103,130	21	1.4	367,193	77,111
cauliflower	47,360	17	1.2	248,712	42,281
celery	26,300	40	1.9	341,391	136,556
cherries– sweet tart	63,220 38,770	10 2	1.2 1.7	281,024 50,703	28,102 1,014
corn (sweet)	271,700	1	2.9	408,706	4,087
cotton (upland)	15,365,000	1	1.6	4,597,962	45,980
cucumbers– fresh processing	56,600 108,210	11 7	4.3 2.9	218,405 164,956	24,024 11,547
eggplant	6,640	15	12.8	48,787	7,318
grapes–all table raisin wine	977,970	10 25 3 11	1.4 1.5 2.5 1.2	2,794,241	279,424
greens– collards kale mustard turnip	14,300 4,880 10,650 11,750	56 40 34 21	5.6 1.6 4.3 3.8	36,217 25,943 51,641 26,795	20,282 10,377 17,558 5,627
lettuce– head other	185,200 99,400	12 6	1.1 1.6	1,208,306 663,204	144,997 39,792
melons (honeydew)	26,200	19	1.7	96,181	18,274
nectarines	36,500	23	1.5	127,642	29,358
peaches	151,820	12	1.5	494,944	59,513
pears	64,630	4	1.5	290,155	11,606
peppers (bell)	64,100	42	4.7	527,452	221,530
plums	37,000	15	1.7	66,443	9,966
prunes	86,000	3	1.2	101,080	3,032

TABLE 2 (continued)

Crops[b]	Acres planted[c]	Percentage acres treated with Bt	Applications per treated acre	Total value of utilized crop ($1000)	Value of crop treated with Bt ($1000)
prunes	86,000	3	1.2	101,080	3,032
raspberries	15,700	38	2.6	92,820	35,272
spinach	32,210	25	1.5	177,585	44,396
squash	56,800	13	3.0	210,287	27,337
strawberries	46,100	41	6.8	1,085,405	445,016
tomatoes–					
fresh	126,100	40	4.6	1,159,590	463,836
processing	309,300	7	1.2	649,066	45,435
watermelon	188,560	11	2.6	241,101	26,521

[a] All data for vegetable, melons and field crops are from 2000 and for fruit crops from 2001. Data for almonds are from 1999. All vegetable and melon acreage refers to acres planted. Acreage for fruits is variable between bearing acreage and acreage harvested.

[b] Sugar beets and oranges were not included in the crop list because less than 1% of the total crop acreage was treated with microbial Bt. Pecans and bulb onions were also not included due to insufficient data.

[c] All vegetable and melon acreage refers to acres planted. Acreage for fruits is variable between bearing acreage and acreage harvested.

Data sources:
Crop acreage and crop values:
–Fruits in 2001 and almonds in 1999 (USDA-NASS 2002a).
–Vegetables and melons in 2000 (USDA-NASS 2002b).
–Cotton in 2000 (USDA-NASS 2001a; USDA-NASS 2001b).
Percentage of acres treated with microbial Bt and average number of Bt applications: (USDA-NASS 2002c).

applied to about 40% of raspberries and strawberries, crops that tend to bear fruit over a long period of time, requiring frequent farm worker entry into fields and repeated harvesting. At the same time, several lepidopteran pests may cause damage to these high-value crops if not controlled during the harvesting season. Most microbial Bt agricultural products may be applied up to the day of harvest and farm workers may re-enter fields 4 hours after Bt applications because of the low human toxicity of these products (U.S. EPA 1995). The unique niche occupied by microbial Bt products may be best understood by closer examination of their usage on particular crops. The following case studies of microbial Bt use on apples and tomatoes illustrate both the importance of microbial Bt for IPM and resistance management as well as some of the limitations of these products.

Apples

Apples are a high-value, perennial fruit crop grown in at least 35 states. In 2001, the total bearing acreage of about 431,000 acres produced 4.7 million tons of apples worth $1.5 billion (USDA-NASS 2002a). While Washington is the largest producer with over 168,000 bearing acres, California, Michigan, New York, Pennsylvania and Virginia are also significant producers, each with 15,000 or more bearing acres. At least ten states have well-developed IPM programs for apples that include reduced and/or more selective pesticide applications to control lepidopteran pests while preventing pesticide-induced outbreaks of mite pests (Whalon & Croft 1984). Selective and safe pest control products such as Bt are important for apple IPM. As microbial Bt is acceptable under certified organic production, it is also an essential tool for organic apple production, an expanding approach practiced on over 8,800 bearing acres in 1997 (Greene 2001).

Lepidopteran insects are among the key pests causing direct damage to apples in all growing regions, but the relative importance of the different lepidopteran species varies between regions, as does the use of microbial Bt applications. Table 3 describes apple production and direct lepidopteran pest control among certain high-producer states. Organophosphate insecticides such as chlorpyrifos and azinphosmethyl, which have toxicity to non-target wildlife and humans, are the most common materials used for control. Biologically derived pest control products include microbial Bt and pheromone-mediated mating disruption. Microbial Bt is more widely used where the obliquebanded leafroller or OBLR (*Choristoneura rosaceana*) is a primary pest. Two factors may influence the use of microbial Bt for OBLR. One is the widespread development of resistance to organophosphate and pyrethroid insecticides in many North American OBLR populations (e.g., Pree et al. 2001; Lawson et al. 1997; Carrière et al. 1994). Recent observations of cross-resistance between some heavily-used organophosphates and new insect growth regulators (Smirle et al. 2002) suggest that growers may have limited options for OBLR management, potentially increasing the importance of microbial Bt. The other factor that may support use of microbial Bt is the adoption of mating disruption to control another key lepidopteran pest, codling moth (*Cydia pomonella*). Although codling moth is physiologically susceptible to many Bt ICPs, its feeding habit of tunneling deep into fruit makes it extremely difficult to control the pest with microbial Bt. Traditionally, growers in areas of high codling moth activity tended to spray organophosphate insecticides which also control leafrollers. However, the development of codling moth sex pheromone products that disrupt mating behavior of the pest may permit significant reduction in the use of organophosphates. Adoption of mating disruption is high in some western states such as Washington,

TABLE 3. Use of Microbial Bt for Apple Pest Management in Four States

State	Bearing acres in 2001[a]	Percentage acres treated with Bt	Primary lepidopteran pests– control measures[b]	Secondary lepidopteran pests[c]– control measures
Michigan	44,500	14	**Leafroller (obliquebanded)** – Chemical: OP, P – Bio: Bt, spinosad, IGR **Codling moth** – Chemical: OP – Bio: mating disruption	**Oriental fruit moth** – Chemical: OP – Bio: mating disruption
New York	55,000	30	**Leafroller (obliquebanded)** – Chemical: OP, P – Bio: Bt, spinosad	Codling moth – Chemical: OP (non-targeted) – Bio: Bt **European corn borer** – Chemical: OP (non-targeted)
Virginia	15,000	< 1	**Leafroller (tufted apple budmoth)** – Chemical: OP, P – Bio: spinosad	**Codling moth** – Chemical: OP (non-targeted) **Leafroller (redbanded)** – Chemical: OP (usually non-targeted)
Washington	168,000	12	**Codling moth** – Chemical: OP – Bio: mating disruption	**Leafrollers (obliquebanded, pandemis)** – Chemical: OP – Bio: Bt, spinosad

[a] Source: USDA-NASS 2002a.

[b] Pest controls: Chemical: OP = organophosphate insecticide applications, C= carbamate insecticides, P = pyrethroid insecticides; Bio = biorational pest controls including microbial controls, mating disruption and insect growth regulators.

Sources: USDA Crop Profiles: New York 2001; Virginia 2000; Washington 2001: See web address: *http://www.pestdata.ncsu.edu/cropprofiles/cplist.cfm?org=state*. Michigan–Epstein et al. 2002.

[c] Secondary pests are those for which control measures are needed only sporadically or which are generally controlled by pesticide applications targeting other pests.

where about 30% of the total apple and pear acreage was treated with mating disruption for codling moth control (Brunner et al. 2002). With the reduction of organophosphate applications, some growers experience increased pest pressure from leafrollers (Walker & Welter 2001) and turn to microbial Bt as an IPM-compatible control. Microbial Bt is especially appropriate for leafroller control in conjunction with the codling moth mating disruption program because Bt does not precipitate mite outbreaks and, applied once or twice per season, is unlikely to select for resistance.

Transgenic apple plants with a Bt ICP gene have already been developed (Cheng et al. 1994) and commercialized may be sought in the near future (James 1999). Bt-transgenic apples could offer growers a cost-effective way to use Bt to control both codling moth and many leafroller pests. However, the development of Bt resistance could have serious consequences not just for transgenic apples, but for apple IPM in general as well as organic production. Existing insecticide resistance management approaches may be inappropriate for a high-value, perennial crop.

Tomatoes

Tomatoes are grown as two distinct crops, fresh-market and processing. In 2001, over 20 states produce fresh-market tomatoes worth a total of $1.1 billion. The leading producers are Florida and California, which harvested 44,500 and 41,000 acres, respectively (USDA-NASS 2002b). Production of processing tomatoes is dominated by California alone, which harvested 254,000 acres in 2001 (USDA-NASS 2002b). A number of lepidopteran insects are key pests on tomatoes in both Florida and California, including various armyworms, the tomato fruitworm (*Helicoverpa zea*–also a major pest in corn and cotton), and the tomato pinworm, *Keiferia lycopericella*. The main armyworm pest in California is the beet armyworm, *Spodoptera exigua*, while the southern armyworm, *S. eridania*, and the yellow striped armyworm, *S. ornithogalli*, are key pests in Florida. In both states, microbial Bt is commonly used to control armyworms as well as secondary pests such as tomato fruitworm (*Helicoverpa zea*), primarily on fresh-market tomatoes (40% of acres) and to a lesser extent on processing (7% of acres) (USDA-NASS 2002b). Other common controls for armyworms include organophosphates such as methomyl and pyrethroids such as esfenvalerate, but these products tend to precipitate outbreaks of *Liriomyza* spp. leafminers. Tomato pinworm may be controlled through preservation of natural enemies, pheromone mating-disruption or, as a last resort, organophosphate or pyrethroid insecticides (UCIPM 1998).

In Florida tomatoes, the use of microbial Bt, the most commonly applied insecticide, is directly linked to the development and adoption of IPM (Frantz & Mellinger 1998). In the mid-1970s, the broad-spectrum insecticides used to control armyworms and other pests precipitated dramatic outbreaks of leafminers, pests previously kept under control by natural enemies. Growers made as many as 34 insecticide applications per season to control pests (Bloem & Mizell 2000). The tomato IPM programs created in response include the use of microbial Bt for armyworm control as well as monitoring and action thresholds for many insect and disease pests (Frantz & Mellinger 1998). Well-developed and widely adopted tomato IPM programs that include use of microbial Bt exist in California as well as Florida (Vandeman et al. 1994). The long-term efficacy of Bt against armyworms is critical to tomato IPM programs.

Use in Organic Agriculture

A number of Bt formulations are acceptable for certified organic production and are used primarily to control lepidopteran pests. For over a decade, certified organic production has been a small but growing sector of U.S. agriculture. Between 1992 and 1997, certified organic cropland more than doubled in the United States (Greene 2001). By 2001, over 1.3 million acres of cropland were under organic certification (USDA-ERS 2002). Organic acreage has increased particularly in specialty crops such as lettuce (5% of total acreage by 2001) and apples (3% of total acreage), but remains much less significant among large commodity crops (USDA-ERS 2002). The use of microbial Bt is expected to grow with the expansion of certified organic acreage.

Use in Public Health

Bt products, specifically those containing the *israelensis* strain, or Bti, play an important role in mosquito control in the United States. Microbial Bti products are also used in developing countries to control mosquito and blackfly vectors of serious diseases such as malaria and onchocerciasis (Das & Amalraj 1997; Kurtak et al. 1987). The Bti spores are typically applied to aquatic habitats in order to kill mosquito larvae. Bti has been shown in laboratory tests to be moderately toxic to the non-target aquatic organism *Daphnia*, but no such effects are expected in the field when Bti products are applied at the labelled use rate (U.S. EPA 1998b). Microbial products are considered among the least environmentally disruptive larvicides available (FCCMC 1998).

The spread of West Nile virus–thought to be vectored primarily by mosquitoes in the genus Culex in the United States–has caused increased interest in mosquito abatement with microbial insecticides. The product of choice is *Bacillus sphaericus* (Bsp), which is generally more effective than Bti in polluted water (de Barjac & Sutherland 1990), typical breeding sites of Culex mosquitoes (Harwood & James 1979). Usually applied as a microbial product consisting of ICPs and live spores, Bsp may maintain a bacterial population at the application site, increasing its residual effect; the residual effect of *Bacillus sphaericus* in the environment can range as high as 7-9 months after applications in tree holes and roadside ditches (Hertlein et al. 1979; Singer 1980).

Use in Forestry

In U.S. forest insect control programs, Bt var. *kurstaki* is used mainly against gypsy moth, *Lymantria dispar*, which currently infests all or parts of 16 northeastern states and can defoliate millions of acres of hardwood forests each year (Liebhold et al. 1997). Since 1980, 1.7 million hectares (4.2 million acres) of eastern U.S. forests have been treated with microbial Bt as part of the Federal, State and County Gypsy Moth Cooperative Suppression Program (Reardon et al. 1999). Over the past decade, however, the forestry market for Bt has declined due to several factors. Unlike more localized agricultural applications of Bt, large-scale forest applications may have at least short-term impacts on non-target lepidoptera (Miller 1990). In 1992, the USDA Forest Service began a program to test the feasibility of more reduced and targeted Bt applications to slow the spread of the gypsy moth in North America. Implementation has resulted in a 60% reduction in the rate of spread (Sharov & Liebhold 1998).

COMPARISON OF MICROBIAL
AND PLANT-INCORPORATED Bt INSECTICIDES

The development and commercialization of transgenic crops that express one or more incorporated Bt ICP genes has greatly expanded the use of Bt-derived insecticides in agriculture, but also raises many questions about the future roles and importance of microbial versus plant-incorporated forms of Bt. As closely related products, microbial and plant-incorporated Bt both have not been found to exhibit significant toxicity to mammals, and are generally considered non-disruptive to beneficial insects. Research in still ongoing to assess non-target effects of plant-incorporated Bt crops. However, plant-incorporated and microbial Bt products function differently in field applications and exhibit distinct advantages and disadvantages relative to each other. Table 4 summarizes these differences.

One of the most important issues in comparing the two types of Bt products is the potential for developing insect resistance to Bt ICPs. While numerous chemical insecticides have lost their effectiveness over years of use, microbial Bt products have continued to demonstrate

TABLE 4. Comparison of Microbial and Plant-Incorporated Bt Products in Agricultural Applications

	Microbial Bt	Plant-incorporated Bt
Ease of application	Disadvantage: Microbial Bt must be applied using conventional equipment, taking care to assure good coverage. Often multiple applications are required.	Advantage: No applications are required. Reduced need for scouting. High, uniform doses of Bt are produced continuously by the plant (under good growing conditions).
Flexibility of application	Advantage: Timing, dosage and Bt formulation can be controlled in any growing season to meet specific pest pressures.	Disadvantage: Bt ICP expression cannot (at this point) be changed once the crop is planted. Only one or two ICPs present, while synergistic compounds found in microbial formulations are absent. Growers must invest in Bt crop before pest pressure is determined. Limited choices of crop varieties may be available.
Efficacy against target pests	Disadvantage: Sprays are not effective against burrowing or soil-dwelling pests. Sprays are not always effective against older larvae.	Advantage: High levels of ICPs can be delivered to all plant parts vulnerable to damage by physiologically susceptible pests.
Environmental persistence	Advantage: Microbial Bt ICPs typically break down rapidly in the phyllosphere as a result of exposure to UV light. Microbial Bt may, however, persist in soil for several months (U.S. EPA 1998b).	Possible disadvantages: Some evidence suggests that soil may bind active Bt ICPs for extended periods of time (Crecchio & Stotsky 1998), but there are no known toxicological effects on non-target soil organisms (Nester et al. 2002; U.S. EPA 2001). Gene transfer between closely related species of plants may occur via pollen transfer from transgenic plants to sexually-compatible wild relatives.
Potential for resistance development	Advantage: The presence of multiple toxins and the short exposure periods reduce selection for resistant target pests (although resistance can still develop, particularly if multiple applications are used).	Disadvantages: The potential for development of Bt resistance by the target insects may be greater due to high dose and season-long expression. Growers must invest in implementing insect resistance management requirements. Advantage: The risk of Bt resistance is mitigated through the implementation of required insect resistance management strategies.
Cost	Disadvantage: Microbial Bt applications can be expensive, particularly if multiple applications are necessary.	Advantage: Bt transgenic crops may be cost-effective if pest pressure threatens to be more costly than the technology.
Market issues	Advantage: Some microbial Bt products are acceptable in certified organic production.	Disavantage: Bt transgenic crop varieties are not approved for organic production. Regulatory and public acceptance of transgenic plants is variable throughout the world.

effective control against targeted pests. The diamondback moth is the only insect species that has developed high levels of resistance to Bt in the field, and the Indian meal moth has developed resistance to Bt in stored grain (Ferré & Van Rie 2002). Bt-resistant strains of other lepidoptera such as *Heliothis virescens* and *Plodia interpunctella* and the coleopteran *Leptinotarsa decemlineata* have been created experimentally through laboratory selection, indicating that many pest species could develop resistance to Bt (Tabashnik 1994). Plant-incorporated Bt crops have the potential to speed up the development of insect resistance to both plant-incorporated and microbial Bt. Therefore, a closer examination of Bt resistance management strategies is useful to better understand the relative merits and predict the future roles of the two types of Bt products.

Bt Resistance Management

The management of potential Bt resistance has become a primary mission with the introduction and expanded use of transgenic plants where Bt ICPs have been genetically engineered into plants for insect control. Tabashnik et al. (1998) point out "With laboratory-selected resistance to Bt demonstrated in many pests and field-evolved resistance to Bt documented in the diamondback moth, adaptation by pests is now considered the biggest threat to the long-term success of transgenic Bt."

Because of the increased concern about insect resistance to Bt expressed in transgenic crops (Bt crops), EPA has mandated specific IRM programs to protect not only the benefits of Bt crops, but also the benefits of microbial Bt formulations (U.S. EPA 2001; Matten et al. 2002) (see Matten & Reynolds, this volume, pp. 137-179). The required IRM programs for Bt crops are both proactive and unprecedented in detail. They include the use of refuges, which are specific plantings of a non-Bt crop intended to support susceptible insects in that can randomly mate with resistant insects to dilute resistance. In addition, EPA has mandated annual resistance monitoring, remedial action plans should resistance develop, and insect resistance management research to improve existing IRM strategies. This type of EPA-mandated IRM program is not required for Bt spray formulations or other conventional insecticides, although EPA has instituted a voluntary resistance management labeling program based on rotation of mode of action (U.S. EPA 2001). There have been no documented reports of insect resistance to Bt ICPs expressed in corn or cotton in the seven years of commercial use of these Bt crops in the U.S. In 2001, approximately 18% of the total U.S.

corn acreage and 37% of the total cotton acreage was planted in Bt varieties (USDA-NASS 2001a). In spite of the high selection intensity, one might attribute the lack of insect resistance to Bt crops to the specific institution of mandated IRM by EPA and the support and adoption by industry and the growers.

Current commercial Bt hybrids contain only single Bt ICP genes. However, the first Bt transgenic crop that contains multiple ICPs (e.g., Bollgard II® cotton) is now commercially available. This concept is known as gene pyramiding, the intent of which is to improve insect control and act as additional insurance against resistance. This strategy relies on each ICP having a unique mode of action, the same target spectrum, and a high dose. The concept of gene pyramiding is partially based on the primary advantage of microbial Bt formulations, which contain multiple ICPs and other toxins, and may offer additional insurance against the development of insect resistance. Under high selection intensities, despite multiple ICPs and other toxins, resistance to sprayable Bts is also of concern, as was observed with diamondback moth (Roush 1994).

The high dose/structured refuge strategy (U.S. EPA 2001; Matten et al. 2002) (see Matten & Reynolds, this volume, pp. 137-178) may not be appropriate for all Bt crop-pest systems. Some Bt crops may need to use other IRM strategies that do not involve a high dose. The use of structured refuges is only one insect resistance management tactic for Bt crops. Other tactics might include seed mixes, temporal plantings of Bt and non-Bt crops, tissue-specific expression, and gene stacking (see Sharma et al., this volume, pp. 53-76). Mating attractants may be used to enhance random mating. Releases of sterile insects in geographically isolated areas may lead to suppression or eradication of resistant insects (e.g., pink bollworm area-wide suppression/eradication programs) (U.S. EPA 2001).

As more Bt crops are introduced into the environment, different insect resistance management strategies will be needed to manage resistance to a wide range of insects. These insects will differ in their biology, ecology, population dynamics and genetics. Highly mobile, polyphagous insect pests, such as *Helicoverpa zea* (Boddie) (known as the corn earworm, cotton bollworm, etc.), may be especially prone to developing resistance if the same or similar Bt ICPs are used in a number of the pest's preferred host crops. There is no "one size fits all" insect resistance management for Bt crops.

CONCLUSIONS

Bt products have a long history of safe use. The industry that markets microbial Bt products has experienced major changes due to the introduction of plant-incorporated Bt products in 1996. However, despite these changes and loss in market share, microbial Bt formulations will continue to play a key role in IPM for fruit and vegetable crops, in sustainable disease vector control and, to a lesser extent, in forestry. In agriculture, plant-incorporated Bt may provide valuable new uses for Bt ICPs, but the benefits of such technologies will only last as long as the target pests remain susceptible. Microbial Bt formulations may be the better option in many cropping systems where the sprays are cost-effective. However, plant-incorporated Bt products may be advantageous in certain situations, particularly on crops previously treated only with broad-spectrum, chemical insecticides. Both types of Bt products should be used in conjunction with IPM programs that involve resistance management. Bt is an important agricultural tool, worthy of extensive effort to preserve its effectiveness.

REFERENCES

Aronson, A.I. and Y. Shai (2001) Why *Bacillus thuringiensis* insecticidal toxins are so effective: unique features of their mode of action. *FEMS Microbiology Letters* 195:1-8.

Berliner, E. (1915) Uber die Schlaffsucht der Mehlmottenraupe (*Ephestia kuhniella* Zell.) und ihren Erreger *Bacillus thuringiensis* n.sp. *Zeitschrift fur Angewandte Entomologie* 2:29-56.

Bloem, S. and R.F. Mizell (2000) Integrated pest management and Florida tomatoes: A success story in progress. University of Florida, Institute of Food and Agricultural Sciences, Extension Brochure. Web address: <http://nfrec-sv.ifas.ufl.edu/ipm_tomato_rpt.htm>.

Brunner, J., S. Welter, C. Calkins, R. Hilton, E. Beers, J. Dunley, T. Unruh, A. Knight, R. Van Steenwyk and P. Van Buskirk (2002) Mating disruption of codling moth: a perspective from the Western United States. IN: Pheromones and other semiochemicals in integrated control. International Organization for Biological and Integrated Control of Noxious Animals and Plants. WPRS Bulletin 25:11-20.

Carriere, Y., J.P. Deland, D.A. Roff and C. Vincent (1994) Life history costs associated with the evolution of insecticide resistance. *Proceedings of the Royal Society of London*, B258:35-40.

Cheng, J.S., E.C.S. Tian, Y.C. Meng and X.M. Mang (1994) Regeneration of transgenic apple plants with the insect resistance gene Bt. *China Fruits* 4:14-15.

Crecchio, C. and G. Stotzky (1998) Insecticidal activity and biodegradation of the toxin from *Bacillus thuringiensis* subsp. *kurstaki* bound to humic acids from soil. *Soil Biology and Biochemistry* 30:463-470.

Croft, B.A. (1990) *Arthropod Biological Control Agents and Pesticides.* John Wiley & Sons, New York.

Das, P.K. and D.D. Amalraj (1997) Biological control of malaria vectors. *Indian Journal of Medical Research* 106:174-197.

de Barjac, H. and D.J. Sutherland (1990) *Bacterial Control of Mosquitoes and Black Flies: Biochemistry, Genetics, and Applications of Bacillus thuringiensis israelensis and Bacillus sphaericus.* Rutgers University Press, New Jersey.

DeBach, P. and D. Rosen (1990) *Biological Control by Natural Enemies.* Cambridge University Press, Cambridge, UK.

Dent, D.R. (1993) The use of *Bacillus thuringiensis* as an insecticide. IN: Jones, D.G. (ed.), *Exploitation of Microorganisms.* Chapman & Hall, Cambridge, UK, pp. 19-44.

Dulmage, H.T. (1970) Insecticidal activity of HD-1, a new isolate of *Bacillus thuringiensis* var. *alesti. Journal of Invertebrate Pathology* 15:232-239.

Dulmage, H.T. and K. Aizawa (1982) Distribution of *Bacillus thuringiensis* in nature. IN: Kurstak, E. (ed), *Microbial and Viral Pesticides.* Marcel Dekker, New York, pp. 209-237.

Epstein, D., D. Waldstein and C.E. Edson (2002) Michigan Apple Integrated Pest Management Implementation Project, Final Narrative Report. Web address: <http://www.cips.msu.edu/maipmip/MAIPMIPFinalReport.pdf>.

Ferré, J. and J. Van Rie (2002) Biochemistry and genetics of insect resistance to *Bacillus thuringiensis. Annual Review of Entomology* 47:501-533.

Flexner, J.L., B. Lighthart and B.A. Croft (1986) The effects of microbial pesticides on non-target, beneficial arthropods. *Agriculture, Ecosystems and Environment* 16: 203-254.

Florida Coordinating Council on Mosquito Control (FCCMC) (1998) Florida mosquito control: the State of the Mission as defined by mosquito controllers, regulators and environmental managers. University of Florida, USA.

Frantz, G. and H.C. Mellinger (1998) Measuring Integrated Pest Management adoption in south Florida vegetable crops. *Proceedings of the Florida State Horticultural Society* Paper No. 132.

Goldberg, L. and J. Margalit (1977) Bacterial spore demonstrating rapid larvicidal activity against *Anopheles sergentii, Uranotaenia unguiculata, Culex univittatus, Aedes aegypti,* and *Culex pipiens. Mosquito News* 37:355-358.

Greene, C.R. (2001) U.S. organic farming emerges in the 1990s: Adoption of certified systems. U.S. Department of Agriculture, Economic Research Service, Resource Economics Division, Agriculture Information Bulletin No. 770.

Harwood, R.F. and M.T. James (1979) *Entomology in Human and Animal Health.* Macmillan Publishing Company Inc., New York.

Hertlein, B.C., R. Levy and T.W. Miller (1984) Persistent spores and mosquito larvicidal activity of *Bacillus sphaericus* 1593 in well water and sewage. *Journal of Invertebrate Pathology* 33: 217-221.

James, C. (1999) Global status of commercialized transgenic crops: 1999. International Service for the Acquisition of Agri-Biotech Applications (ISAAA) Briefs No. 17-2000.

James, R.R., J.C. Miller and B. Lighthart (1993) *Bacillus thuringiensis* var. *kurstaki* affects a beneficial insect, the cinnabar moth (Lepidoptera: Arctiidae). *Journal of Economic Entomology* 86:334-339.

Knowles, B.H. (1994) Mechanism of action of *Bacillus thuringiensis* insecticidal endotoxins. *Adv. Insect Physiol.* 24:275-308.

Krieg, V.A., A.M. Huger, G.A. Langerbruch and W. Schnetter (1983) *Bacillus thuringiensis* var. *tenebrionis*: ein neuer, gegenuber Larven von Coleoptera wirksamer Pathotyp. *Zeitschrift fur Angewandte Entomologie* 96:500-508.

Kurtak, D., H. Jamnback, R. Meyer, M. Ocran and P. Renaud (1987) Evaluation of larvicides for the control of *Simulium damnosum* s.l. (Diptera: Simuliidae) in West Africa. *Journal of the American Mosquito Control Association* 3:201-210.

Lawson, D.S., W.H. Reissig and C.M. Smith (1997) Response of larval and adult obliquebanded leafroller (Lepidoptera: Tortricidae) to selected insecticides. *Journal of Economic Entomology* 90:1450-1457.

Liebhold, A.M., K.W. Gottschalk, E.R. Luzader, D.A. Mason, R. Bush and D.B. Twardus (1997) *Gypsy moth in the United States: an Atlas*. USDA Forest Service, Northeastern Forest Experiment Station, General Technical Report NE-233.

Lüthy, P., J. Cordier and H. Fischer (1982) *Bacillus thuringiensis* as a bacterial insecticide: basic considerations and application. IN: Kurstak, E. (ed.), *Microbial and Viral Pesticides*. E. Marcel Dekker Publishers, New York, pp. 35-74.

Matten, S.R., R. Hellmich and A.H. Reynolds (2003) Current resistance management strategies for Bt corn in the United States. IN: Koul, O., Dhaliwal, G.S. (eds.), *Transgenic Crop Production: Concepts and Strategies*. Science Publishers, Inc., U.S.A., In press.

Miller, J.C. (1990) Field assessment of the effects of a microbial pest control agent on nontarget lepidoptera. *American Entomologist* 36:135-139.

Nester, E., L. Thomashow, M. Metz and M. Gordon (2002) 100 Years of *Bacillus thuringiensis*: A Critical Scientific Assessment. A Report from the American Academy of Microbiology, Washington, DC.

Pree, D.J., K.J. Whitty, M.K. Pogoda and L.A. Bittner (2001) Occurrence of resistance to insecticides in populations of the obliquebanded learfoller from orchards. *Canadian Entomologist* 133:93-103.

Reardon, R., N. DuBois and W. McLane (1999) *Bacillus thuringiensis* for managing gypsy moth: a review. Forest Service, U.S. Department of Agriculture. *Gypsy Moth News* 47:2.

Roush, R.T. (1994) Managing pests and their resistance to *Bacillus thuringiensis*: Can transgenic crops be better than sprays? *Biocontrol Science and Technology* 4:501-516.

Sharov, A.A. and A.M. Liebhold (1998) Model of slowing the spread of gypsy moth (Lepidoptera: Lymantriidae) with a barrier zone. *Ecological Applications* 8:1170-1179.

Singer, S. (1980) *Bacillus sphaericus* for the control of mosquitoes. *Biotechnology and Bioengineering* 22:1335-1355.

Smirle, M.J., D.T. Lowery and C.L. Zurowski (2002) Resistance and cross-resistance to four insecticides in populations of obliquebanded learoller (Lepidoptera: Tortricidae). *Journal of Economic Entomology* 95:820-825.

Tabashnik, B.E. (1994) Evolution of resistance to *Bacillus thuringiensis*. *Annual Review of Entomology* 39:47-79.
Tabashnik, B.E., T-B. Liu, Y. Malvar, D.G. Heckel, L. Masson and J. Ferré (1998) Insect resistance of *Bacillus thuringiensis*: uniform or diverse? *Phil. Trans. R. Soc. Lond. B. 353*, 1751-1756.
United States Department of Agriculture–Economic Research Service (USDA-ERS). (2002). Organic Production Data. Web Address: <*http://www.ers.usda.gov/Data/organic/*>.
United States Department of Agriculture–National Agricultural Statistics Service (USDA-NASS) (2001a) Crop production–2000 Summary. U.S. Department of Agriculture, Economics and Statistics System. <*http://usda.mannlib.cornell.edu/*>.
USDA-NASS (2001b) Crop values–2000 Summary. U.S. Department of Agriculture, Economics and Statistics System. <*http://usda.mannlib.cornell.edu/*>.
USDA-NASS (2002a) Noncitrus Fruits and Nuts 2001 Preliminary Summary. U.S. Department of Agriculture, Economics and Statistics System. <*http://usda.mannlib.cornell.edu/*>.
USDA-NASS (2002b) Vegetable Summary 2001. U.S. Department of Agriculture, Economics and Statistics System. <*http://usda.mannlib.cornell.edu/*>.
USDA-NASS (2002c) Agrichemical Use Database. Web address: <*http://www.pestmanagement.info/nass/*>.
United States Environmental Protection Agency (U.S. EPA) (1995) Pesticide Registration Notice 95-3. Reduction of worker protection standard (WPS) interim restricted entry intervals (REIs) for certain low risk pesticides. <*http://www.epa.gov/opppmsdl/ PR_Notices/pr95-3.html*>.
United States Environmental Protection Agency (U.S. EPA) (1996) Pesticide Program Dialogue Committee–Meeting Summary, July 9-10, 1996, Arlington, VA. Web address of meeting summary: <*http://www.epa.gov/pesticides/ppdc/july96.htm*>.
U.S. EPA (1998a) White Paper on Bt Plant-Pesticide Resistance Management. U.S. EPA, Biopesticides and Pollution Prevention Division (7511C). 14 January 1998, EPA Publication 79-S-98-001.
U.S. EPA (1998b) *Bacillus thuringiensis* Reregistration Eligibility Document. Office of Pesticide Programs. Web address: <*http://www.epa.gov/oppsrrdl.REDs/0247.pdf*>.
U.S. EPA (1998c) Biopesticide Fact Sheet: *Bacillus thuringiensis* subspecies *israelensis* strain EG2215. Web address: <*http://www.epa.gov/pesticides/biopesticides/factsheets/fs006476t.htm*>.
U.S. EPA (1998d) Biopesticide Fact Sheet: *Bacillus thuringiensis* subspecies *kurstaki* strain M-200. Web address: <*http://www.epa.gov/pesticides/biopesticides/factsheets/fs006452t.htm*>.
U.S. EPA (2001) *Bacillus thuringiensis* Plant-Incorporated Protectants, Biopesticides Registration Action Document. Office of Pesticide Programs, October 16, 2001. Web address: <*http://www.epa.gov/pesticides/biopesticides/reds/brad_bt_pip2.htm*>.
U.S. EPA (Undated) (EPA PR Notice 2001-5). Guidance for Pesticide Registrants on Pesticide Resistance Management Labeling. Web address: <*http://www.epa.gov/opppmsdl/PR_Notices/pr2001-5.pdf*>.
University of California–Statewide IPM Program (UCIPM) (1998) Integrated Pest Management for Tomatoes, 4th Ed. University of California, Agriculture and Natural Resources Division, Publication 3274.

UCIPM (1999) *Integrated Pest Management for Apples and Pears, 2nd Ed.* University of California, Agriculture and Natural Resources Division, Publication 3340.

Vandeman, A., J. Fernandez-Conejo, S. Jans and B.H. Line (1994) Adoption of Integrated Pest Management in U.S. Agriculture. Agricultural Information Bulletin No. 707. Economic Research Service, Washington, DC.

Walker, K. and S. Welter (2001) Potential for outbreaks of leafrollers (Lepidoptera: Tortricidae) in California apple orchards using mating disruption for codling moth suppression. *Journal of Economic Entomology* 94:373-380.

Ware, G.W. (1989) *The Pesticide Book, 3rd ed.* Thomson Publications, Fresno, CA.

Whalon, M.E. and B.A. Croft (1984) Apple IPM implementation in North America. *Annual Review of Entomology* 29:435-470.

The Utility and Management of Transgenic Plants with *Bacillus thuringiensis* Genes for Protection from Pests

Hari C. Sharma
Kiran K. Sharma
Nadoor Seetharama
Jonathan H. Crouch

SUMMARY. Recombinant DNA technology offers opportunities for widening the available gene pool for crop improvement. Genetic engineering also allows the introduction of several desirable genes in a single event, and can reduce the time to introgress novel genes into elite backgrounds. Genes conferring resistance to insects have been inserted into crop plants such as cotton, maize, potato, tobacco, rice, broccoli, lettuce, walnut, apple, alfalfa, and soybean. Genetically transformed crops with *Bacillus thuringiensis* (Bt) genes have been deployed for cultivation primarily in the USA, China, Argentina, Canada, Mexico, South Africa,

Hari C. Sharma, Kiran K. Sharma and Jonathan H. Crouch are affiliated with the International Crops Research Institute for the Semi-Arid Tropics (ICRISAT), Patancheru 502 324, Andhra Pradesh, India.

Nadoor Seetharama is affiliated with the National Research Center for Sorghum, Rajendernagar, Hyderabad 500 030, Andhra Pradesh, India.

Address correspondence to: Hari C. Sharma (E-mail: h.sharma@cgiar.org).

[Haworth co-indexing entry note]: "The Utility and Management of Transgenic Plants with *Bacillus thuringiensis* Genes for Protection from Pests." Sharma, Hari C. et al. Co-published simultaneously in *Journal of New Seeds* (Food Products Press, an imprint of The Haworth Press, Inc.) Vol. 5, No. 1, 2003, pp. 53-76; and: *Bacillus thuringiensis: A Cornerstone of Modern Agriculture* (ed: Matthew Metz) Food Products Press, an imprint of The Haworth Press, Inc., 2003, pp. 53-76. Single or multiple copies of this article are available for a fee from The Haworth Document Delivery Service [1-800-HAWORTH, 9:00 a.m. - 5:00 p.m. (EST). E-mail address: docdelivery@haworthpress.com].

http://www.haworthpress.com/store/product.asp?sku=J153
© 2003 by Taylor & Francis.
10.1300/J153v05n01_04

and Australia. The potential of insect-resistant transgenic plants with Bt genes can be enhanced when deployed in combination with alternate protective genes such as protease inhibitors, enzymes, and plant lectins, or in combination with insect-resistant cultivars derived through conventional breeding. While several transgenic crops with insecticidal genes have been introduced in the temperate regions, very little has been done to use this technology for improving crop production in the harsh environments of the tropics, where the need for increasing food production is most urgent. This may be due to the lack of infrastructure, biosafety regulations, intellectual property rights, or market potential. There is an urgent need to develop a scientifically sound strategy to deploy exotic and plant derived genes through transgenic plants for minimizing the extent of losses caused by insect pests. Equally important is the need for observance of biosafety regulations, a responsible public debate, and a better presentation of the benefits to sustainable crop production of a rational deployment of genetically transformed plants. *[Article copies available for a fee from The Haworth Document Delivery Service: 1-800-HAWORTH. E-mail address: <docdelivery@haworthpress.com> Website: <http://www.HaworthPress. com> © 2003 by The Haworth Press, Inc. All rights reserved.]*

KEYWORDS. Genetically modified plants, *Bacillus thuringiensis*, crop production, resistance management, pest management

INTRODUCTION

There is an urgent need to increase food production, particularly in the developing countries, to feed the rapidly increasing human population. One of the means of increasing food production is to minimize crop losses associated with insect pests, which currently are estimated at 14% of the total agricultural production (Oerke et al., 1994). The additional cost in the form of insecticides applied for controlling insect pests is estimated at US$10 billion annually. Insect control involving chemicals can result in toxic residues in food and food products and adverse effects on non-target organisms and the environment. Therefore, it is important to develop technologies that can minimize the pesticide use in crop production. It is in this context that biotechnology can make a major contribution to pest management (Hilder and Boulter, 1999; Sharma et al., 2002).

Much of the success in developing insect-resistant transgenic plants with *Bacillus thuringiensis* (Bt) genes has come from the in-depth genetic and biochemical information of this bacterium. Bt was first isolated from the diseased larvae of *Ephetia kuhniella* (Berliner, 1915). The insecticidal activity of this gram-positive bacterium is provided by proteinaceous crystalline (Cry) inclusion bodies and cytotoxins. There are several subspecies of this bacterium, which are effective against the lepidopteran, dipteran, and coleopteran insects. The Cry Bt insecticidal crystal proteins (ICPs) were earlier classified into four types, based on insect specificity and sequence homology (Hofte and Whiteley, 1989). Type I *cry* genes encode proteins of 130 kDa, which are usually specific to lepidopteran (butterflies and moths) larvae, type II genes encode for 70 kDa proteins that are specific to lepidopteran and dipteran (flies and mosquitoes) larvae, and type III genes encode for 70 kDa proteins specific to coleopteran (beetles) larvae, while type IV genes are specific to the dipteran larvae. The system was further extended to include type V genes that encode for proteins that are effective against lepidopteran and coleopteran larvae (Tailor et al., 1992). The Bt ICPs are now known to constitute a family of related proteins for which over 140 genes have been described (Crickmore et al., 1998).

MODE OF ACTION OF Bt TOXINS

Bt ICPs function as midgut toxins, and are effective only when ingested by the target insect pests. Insect mortality may occur in hours to days, and takes much longer than for synthetic insecticides. In transgenic crops having Bt genes, the plant tissues produce specific Cry proteins in a soluble form that traverse the peritropic membrane and bind to specific receptors on the insect midgut epithelium, forming pores and leading to loss of the transmembrane potential, cell lysis, leakage of the midgut contents, paralysis, and death of the insect (Gill et al., 1992). Insects that develop resistance to Bt most commonly exhibit decreased or altered receptor binding or even proteolytic inactivation. When applied as a spray, Bt is relatively unstable, and can be washed off by rain or broken down by ultraviolet light, and may need to be reapplied every 2 to 3 days. This instability along with variable and low-dosage residues on the plant can contribute to the emergence of resistant insect populations (Gould, 1998). A strategy to overcome this is provided by higher and more uniform doses that can be obtained in Bt transgenic plants (Gould, 1994).

EFFICACY OF Bt ICPS AGAINST INSECT PESTS IN DIFFERENT CROPS

With the advent of genetic transformation techniques based on recombinant DNA technology, it is now possible to insert genes into the plant genome and develop plants that are protected against target insect pests. Genes encoding Bt ICPs have been cloned since the 1980s (Schnepf and Whiteley, 1981), and genetically modified, insect protected plants were developed in the mid-1980s (Barton et al., 1987; Fischhoff et al., 1987; Vaeck et al., 1987; Hilder and Boulter, 1999; Sharma et al., 2000; Nelson, 2001).

Bt genes in cotton plants are effective against pink bollworm (*Pectinophora gossypiella*) (Wilson et al., 1992), while transgenic cotton, Coker 312 transformed with the *cry1Ac* gene has shown high levels of resistance to cabbage looper (*Trichoplusia ni*), tobacco caterpillar (*Spodoptera exigua*), and cotton bollworm (*Helicoverpa zea*). In an early demonstration of field performance, transgenic cotton suffered 1.1% bollworm damage and yielded 1,460 kg ha^{-1} cottonseed, compared to 12% damage and a yield of 1,050 kg ha^{-1} in the non-transformed, Coker 312 variety (Benedict et al., 1996). In China, transgenic cotton cultivars Shiyuan 321, Zhongmiansuo 19, 3517 and 541 have resulted in up to 96% mortality of cotton bollworm, *Helicoverpa armigera* (Guo et al., 1999). Cropping with transgenic cotton R 93-4 has been estimated to result in a net income of 23 million Yuan (Diao and Xie, 1997). Recent studies on Bt cotton in India (Quaim and Zilberman, 2003) and in China (Huang et al., 2002) have shown that yields can be increased with the built-in pest protection. This has also been observed in Argentina (see deBianconi, this volume, pp. 223-235). Thus, there seems to be a clear advantage of growing transgenic cotton in reducing insect damage and increasing the cottonseed yield.

Transgenic Bt maize is highly effective against the European corn borer (*Ostrinia nubilalis*) (Koziel et al., 1993; Armstrong et al., 1995; Archer et al., 2000). Transgenic maize expressing Cry9C (from *Bacillus thuringiensis* subs. *tolworthi*) is highly effective against the European corn borer (Jansens et al., 1997). Transformed maize plants are also effective against the spotted stem borer (*Chilo partellus*) and the maize stalk borer (*Busseola fusca*) (Rensburg van, 1999). Spotted stem borer is more susceptible than maize stalk borer to transgenic maize with Bt genes. Maize plants with *cry1Ab* gene are also resistant to the sugarcane borers, *Diatraea grandiosella* and *Diatraea saccharalis* (Bergvinson et al., 1997). The Bt-transformed plants exhibit greater resistance to *D. grandiosella* than those derived from conventional host plant resistance breeding programs. Transgenic maize hybrids have been found to suffer significantly less leaf damage by *Spodoptera frugiperda* and Southwestern corn borer (*Diabrotica undecimpuncta howardi*) than the resistant cultivars derived through conventional plant breeding (Williams et al., 1997). Resistance to fall armyworm and near immunity to Southwestern corn borer observed in these transgenic maize hybrids is the highest level of resistance documented for these insect pests. Transgenic tropical maize inbred lines with *cry1Ab* or *cry1Ac* synthetic genes with resistance to corn earworm, fall armyworm, Southwestern corn borer, and sugarcane borer have also been developed (Bohorova et al., 1999). Truncated *cry1Ab* gene in sugarcane has also shown significant activity against the sugarcane borer (*D. saccharalis*) (Arencibia et al., 1997) (see Braga et al., this volume, pp. 209-221).

Transgenic corn has shown considerable promise for reducing insect pest-associated losses, and its effectiveness is enhanced in combination with other components of pest management. Bt maize is quite effective in preventing European corn borer damage and produces higher grain yields (Clark et al., 2000). In the absence of European corn borer infestation, the performance of transgenic hybrids is similar to their nontransgenic counterparts. However, the yield of isoline hybrids is 10% lower than the standard and Bt hybrids regardless of European corn borer infestation (Lauer and Wedberg, 1999), but Bt hybrids generally yield 4 to 8% greater than the standard hybrids when infested with European corn borer. Transgenic hybrids with *cry1Ab* also suffer less *Fusarium* ear rot than their nontransgenic counterparts (Munkvold et al., 1999).

Rice plants having 0.05% ICP of the total soluble leaf protein have shown high levels of resistance to the striped stem borer (*Chilo suppressalis*) and rice leaf folder (*Cnaphalocrosis medinalis*) (Fujimoto et al., 1993). Scented varieties of rice (Basmati 370 and M 7) transformed with *cry2a* are resistant to yellow rice stem borer (*Scirpophaga incertulas*) and the rice leaf folder (Mqbool et al., 1998). Truncated *cry1Ab* gene has been introduced into several *indica* and *japonica* rice cultivars by microprojectile bombardment and protoplast systems (Datta et al., 1998). Rice lines transformed with the synthetic *cry1Ac* gene are also resistant to yellow stem borer larvae (Nayak et al., 1997). Rice plants expressing *cry1Ab* gene are toxic to the striped stem borer (*C. suppressalis*) and the yellow stem borer (Ghareyazie et al., 1997). The *cry1Ab* gene has also been inserted into the maintainer line, R 68899B, with enhanced resistance to yellow stem borer (Alam et al., 1999). A more detailed description of a major effort to engineer Bt ICPs into rice is given by Datta et al. (this volume, pp. 77-91).

A codon-modified *cry1Ac* gene has been introduced into groundnut (peanut) (Singsit et al., 1997), and the transgenic plants have shown resistance to the lesser corn stalk borer (*Elasmopalpus lignosellus*). Chickpea cultivars ICCV 1 and ICCV 6 have been transformed with *cry1Ac* genes, which inhibit the development and feeding of *H. armigera* (Kar et al., 1997).

Expression of *cry1Ac* (Mandaokar et al., 2000) and *cry1Ab* (Shukla et al., 2003) gene in tobacco plants is effective against *H. armigera*. Synthetic *cry3* genes in tobacco are also effective for the control of Colorado potato beetle (*Leptinotarsa decemlineata*) (Perlak et al., 1993). Tobacco plants containing the *cry2a5* gene are highly resistant to *H. armigera* (Selvapandian et al., 1998), and the effectiveness of this ICP gene is comparable to *cry1Ab* or *cry1Ac*. Tomato plants expressing Cry1Ab and Cry1Ac proteins are effective against lepidopteran insects (Delannay et al., 1989; Van der Salm et al., 1994).

Synthetic *cry3* gene has been expressed in potato plants, protection against conferring Colorado potato beetle (*L. decemlineata*) (Jansens et al., 1995). Transgenic potato plants containing the Cry1Ab ICP (Bt 884), and a truncated ICP, Cry1Ab6, are less damaged by potato tuber moth (*Pthorimaea opercullela*) (Arpaia et al., 2000). Transgenic LT 8 and Sangema tubers have shown high levels of resistance to *P. operculella* up to 6 months. No significant effects have been observed on the non-target species such as *Liriomyza huidobrensis*, *Russelliana solanicola* and *Myzus persicae*. Damage to the 4th terminal leaf in transgenic plants by *Epitrix cucumeris* was 20 to 31% lower as compared to non-transgenic plants (Stoger et al., 1999). Cry5-Lemhi Russet and Cry5-Atlantic potato lines have resulted in up to 100% mortality of first-instar larvae of the potato tuber moth (Mohammed et al., 2000). The expression of Cry1Ab in potato cultivars Sangema, Cruza 148 and LT 8 showed up to 100% *P. operculella* larval mortality (Canedo et al., 1999). New Leaf™ Bt-transgenic potatoes provide substantial ecological and economic benefits to potato growers (Hoy, 1999) in the form of reduced insecticide inputs. A more extensive account of work with Bt potato is given by Ghislain et al. (this volume, pp. 93-113).

Transformed brinjal (eggplant) plants have shown significant insecticidal activity against the fruit borer (*Leucinodes orbonalis*) (Kumar et al., 1998). A modified gene of *B. thuringiensis* var. *tolworthi* (*cry3B*) has shown insecticidal activity against Colorado potato beetle (Arpaia et al., 1997). Synthetic *cry1C* gene introduced into broccoli (*Brassica oleracea* subsp. *italica*) provides protection not only from susceptible diamond back moth (*Plutella xylostella*) larvae, but also from diamond back moth selected for moderate levels of resistance to Cry1C protein (Cao et al., 1999). Transgenic broccoli containing Cry1C is also resistant to the cabbage looper (*Trichoplusia ni*) and cabbage butterfly (*Pieris rapae*). *Brassica campestris* subsp. *parachinensis* expressing Cry1Ab or Cry1Ac has shown resistance to *P. xylostella* (Xiang et al., 2000). Recent work by Cao et al. has also shown that Cry1C in transgenic cauliflower can confer protection against diamondback moth and cabbage looper (Cao et al., this volume, pp. 193-207).

DEVELOPMENT OF INSECT-RESISTANT TRANSGENIC PLANTS AT ICRISAT

Insect pests are major constraints to productivity of ICRISATs' mandate crops (sorghum, pearl millet, pigeonpea, chickpea, and groundnut) in the semi-arid tropics. The avoidable losses due to insect pests have been estimated at over US$14 billion annually (ICRISAT, 1992). Efforts are underway at ICRISAT to develop transgenic plants with Bt, protease inhibitors, and lectin genes for resistance to spotted stem borer, *C. partellus* in sorghum, and pod borer, *H. armigera* in pigeonpea and chickpea (Sharma and Ortiz, 2000). Bt ICPs from *B. thuringiensis* var *morrisoni* have shown biological activity against the sorghum shoot fly (*Atherigona soccata*). Cry1Ac and Cry2A ICPs are moderately effective against spotted stem borer (*C. partellus*), while Cry1Ab is effective against *H. armigera* (Sharma et al., 1999). Efficient tissue culture and transformation protocols using microprojectile bombardment have been developed, and sorghum plants expressing Cry1Ac have been developed. These are presently being tested for their resistance to the spotted stem borer (Harshavardhan et al., 2002). In chickpea and pigeonpea, tissue culture and transformation methods using *Agrobacterium tumefaciens* have been standardized (Prasanna et al., 1997; Jayanand et al., 2003; Dayal et al., 2003). Transgenic plants with Bt *cry1Ab* and *cry1Ac* and soybean trypsin inhibitor genes are at different stages of evaluation for resistance to *H. armigera*. Transgenic plants with resistance to the major insect pests in these crops are expected a major role in pest management in the semi-arid tropics (Sharma et al., 2002).

STRATEGIES FOR DEPLOYMENT
OF TRANSGENIC PLANTS

Deployment of transgenic plants requires diligent management if the benefits of biotechnology are to be maintained. Management must take into account alternate mortality factors, reduction of selection pressure, and monitoring of pest populations for resistance development to design effective strategies (Fitt et al., 1994; Gould, 1994, 1998). To prevent incremental breakdown of the technology it is important to implement the resistance management strategies from the beginning of deployment of transgenic, insect protected crops. A number of conceptual strategies have been developed for resistance management (McGaughey and Whalon, 1992; Tabashnik, 1994; Kennedy and Whalon, 1995; Gould, 1998). Most of these strategies are based on mixtures of toxins to be deployed for insect control, tissue specific production, and induced toxin production. Two or more ICP or other anti-insect genes (gene 'pyramiding') can be introduced into the same plant to hamper development of resistance, a strategy pursued by the work of Datta et al. (this volume, pp. 77-91) in rice and experimentally supported by the work of Cao et al. (this volume, pp. 193-207). Plants can also be engineered so that the ICPs are produced only in the tissues where the insect feeds.

Expression of toxins at very high levels can also be used to slow down the adaptation to a toxin if the ecology and genetics of the insect and cropping system fit specific assumptions (Gould, 1994). These assumptions relate to: (i) pattern of inheritance of resistance, (ii) ecological costs of resistance, (iii) behavioral responses of the insect to the toxins, and (iv) distribution of host plants that do, and do not produce the toxin(s). Arpaia et al. (1998) suggested that only a high level of migration (very likely in most agricultural areas) or a sensible reduction of the fitness associated with the change in the genome could guarantee a long-lasting efficacy of the transgenic crops. Increase in summer migration and the distance covered might delay resistance development. It has been suggested that Bt-sunflower might not lead to the development of a Bt-resistant sunflower moth populations (Brewer, 1991). Thus, the strategies for resistance management would depend on the number and nature of gene action, insect behavior, and insect-genotype-environment interaction. To maintain the effectiveness and usefulness of transgenic plants, it is important to develop strategies for resistance management including: (1) gene pyramiding, (2) gene deployment, (3) regulation of gene expression, (4) development of synthetics, (5) refugia, (6) destruction of carryover population, (7) use of planting window, (8) pesticide application, and (9) integrated pest management (IPM).

Gene Pyramiding

Insect populations readily produce resistant individuals when subjected to the selection pressure of chemical insecticides, and should not be expected to do otherwise with toxin proteins deployed through transgenic plants. This concern is acute for many of the candidate genes, that have been used for protection from insects through transgenic crops, because they are either too specific or are only mildly effective against the target insect pests. Therefore, to convert transgenics into an effective weapon in pest control, it is important to deploy genes with different modes of action in the same plant (Zhao et al., 1997). This multi-gene approach is termed gene 'pyramiding' and functions to counter the probability of pests evolving protection from a single protectant gene. The probability of resistance developing simultaneously against multiple genes with different modes of action declines exponentially with an increase in the number of protective genes present in the plant.

Several genes such as trypsin inhibitors, secondary plant metabolites, vegetative insecticidal proteins, plant lectins, and enzymes that are selectively toxic to insects can be deployed along with the Bt genes to increase the durability of transgenic protection from insects (Sharma et al., 2000). Considerable advances have been made in introducing and expressing multiple transgenes in crops (Hadi et al., 1996; Chen et al., 1998). Chakrabarti et al. (1998) suggested that Cry1Ac and Cry1F can be expressed together in transgenic plants for effective control of *H. armigera*. Expression of multiple Bt genes is also presented, with rice treated as a model by Datta et al. (this volume, pp. 77-91), and examined experimentally in cauliflower by Cao et al. (this volume, pp. 193-207). Activity of Bt in transgenic plants can also be enhanced by serine protease inhibitors (MacIntosh et al., 1990), tannic acid (Gibson et al., 1995), and terpenoids (Sachs et al., 1996). Transgenic poplars expressing proteinase inhibitor and *cry3A* genes exhibited reduced larval growth, altered development, and increased mortality as compared to the nontransformed control plants (Cornu et al., 1996) The codon-modified *cry5*-Bt gene, which is specifically toxic to Lepidoptera and Coleoptera, and a potato Y potyvirus Yo coat protein gene (PVYocp), in which the aphid transmission site was inactivated, have been inserted into potato cultivar Spunta using *Agrobacterium tumefaciens* (Li et al., 1999). All Cry5-Bt/PVYocp-transgenic lines have been found to be resistant to potato tuber moth and PVYo infection than the non-transgenic Spunta. Insecticidal action of the transgenic plants expressing both the genes has been found to be significantly higher than that of the plants expressing the Bt gene alone. Only fifth-instar larvae could survive until pupation when fed the Bt + CpTi diet (Zhao et al., 1998).

Gene Deployment

There is a need to develop appropriate strategies for gene deployment in different crops and regions depending on the pest spectrum, their sensitivity to the insecticidal genes, and interaction with the environment. The same Bt gene should not be deployed in crops that serve as an alternate host to the same pest, e.g., *H. armigera* on chickpea, pigeonpea, sorghum, and cotton. The deployment of different genes and their level of expression should be based on insect sensitivity and level of resistance development, e.g., deployment of Cry1Ab in cotton will be less effective in areas where *Spodoptera litura* and *Bemisia tabaci* are also serious pests. Cao et al. (1999) showed that high production of Cry1C protein can protect transgenic broccoli not only from susceptible or Cry1Ab resistant diamondback moth larvae, but also from those selected for moderate levels of resistance of Cry1C. The Cry1C-transgenic broccoli was also resistant to two other lepidopteran pests of crucifers (cabbage looper and imported cabbage worm).

Regulation of Gene Expression

Regulation of gene expression by the use of appropriate promoters is important for durability and specificity of resistance. In most cases, resistance genes have been controlled by constitutive promoters such as cauliflower mosaic virus 35S (*CaMV35S*), maize *ubiquitin* or rice *actin1*, which direct expression in most plant tissues. Limiting the time and place of gene expression by tissue specific promoters such as phenylalanine ammonia lyase (*PHA-L*) for seed specific expression, *RsS1* for phloem specific expression, or inducible promoters such as potato *pin2* wound-induced promoter might contribute to the resistance management, and avoid unfavorable interactions with the beneficial insects. Use of tissue specific pith and PEPC promoters in transgenic rice is presented by Datta et al. (this volume, pp. 77-91) and has been examined experimentally in sugarcane by Braga et al. (this volume, pp. 209-222).

For efficient pest control, it is important that insect control proteins are expressed in adequate amounts at the site where the insects feed. Restricted expression in tissues may also contribute to minimizing the yield penalty associated with the transgene expression (Xu et al., 1993; Schuler et al., 1998). There are specific situations where specific promoters would have a clear advantage such as root feeding insects.

Development of resistance in insects can also be dramatically reduced through the genetic engineering of chloroplasts in plants (Kota et al., 1999). Transformed tobacco leaves expressing Cry2Aa2 ICP at 2 to 3% of the total soluble protein (20- to 30-fold higher than the transgenic plants produced through tissue culture and *Agrobacterium*-mediated transformation) are effective against the resistant populations of *H. zea*, *H. viresens*, and *S. exigua*.

Development of Synthetics

One of the strategies for deploying transgenic plants is the constitutive expression of different Cry toxins in different lines, which can be used to form synthetics (Bergvinson et al., 1997). Synthetics are mixtures of lines which are similar to each other phenotypically, but are diverse in terms of genes conferring resistance to different biotypes of the same pest or to different pests. Once released to the farmers, the synthetics can be maintained as narrow based populations at the farm level by removing the plants showing insect damage. This is an adaptation of traditional selective breeding practices used by farmers throughout history, where seeds from the best plants were used for the next generation/season of plantings. Lines with pest resistance derived through conventional plant breeding can also be included as a component in developing the synthetics to increase the durability of resistance. Pyramiding Cry1Ab insecticidal protein with high-terpenoid content in cotton also increases the level of resistance to *H. virescens* (Sachs et al., 1996). High levels of expression of Cry5 in lines with natural resistance to *P. opercullela* resulted in 96% mortality of this pest (Westedt et al., 1998). Transgenics and lines derived through conventional breeding with different mechanisms of protection can be combined to achieve durable protection against the difficult to control pests (Westedt et al., 1998).

Refugia

One of the main strategies to manage the development of resistance to Bt toxins is using high dose and production of refugia, in which certain percentage of the crop consists of non-Bt plants (4 to 20% in maize, and 20 to 40% in cotton) (DuRant et al., 1996; Gould 1998). The non-Bt plants support a population of susceptible insects, which have a high probability of mating with rare Bt resistant insects emerging from the Bt crops nearby, and thus dilute the frequency of resistant genotypes. The refuges can be sprayed or unsprayed. For refugia to be effective, they should be closer to the transgenic plants, so that insects emerging from the non-transgenic crop have the opportunity to mate with insects produced on the transgenic plants. Sophisticated guidelines for use of refugia in insect resistance management have been developed by the Environment Protection Agency in USA, and are presented by Matten and Reynolds (this volume, pp. 137-178).

For polyphagous pests such as *H. armigera*, which feeds on several crops and alternate hosts in the wild, there may not be any need to maintain the refugia under mixed cropping in the tropics, where several non-transgenic crops may be grown by the farmers in the vicinity of transgenic crop. Separate refuges are superior to seed mixtures for delaying development of resistance. Movement of *H. zea* larvae from non-transgenic to the transgenic plants may result in an increase in damage and reduce the yield in mixed stands of Bt and non-Bt plants (DuRant et al., 1996). The number of eggs of *H. virescens* and *H zea* did not differ between mixed and pure stands of transgenic or non-transgenic cotton plants (Halcomb et al., 1996; Lambert et al., 1996). Transgenic and nontransgenic plants could be grown in strips to minimize the rate of resistance development (Ramachandran et al., 1998; Onstad et al., 1998).

Destruction of Carryover Population

Destruction of pupae or the carryover population (that has been exposed to Bt crops in the previous generations) from one season to another is another important component of resistance management (Fitt and Forrester, 1988; Fitt, 1989; Murray and Zalucki, 1990). Ploughing or flooding the fields immediately after the crop harvest will expose the hibernating larvae or pupae, reducing their viability (Yang et al., 1999). Destruction of stems or burning of stubbles of crops with insect larvae and pupae will also help in reducing the carryover of insects from one season to another (Sharma, 1985). Appropriate crop rotations and observing a "close-season" also reduces the population carryover of several insect species (Sharma, 1993). Removal of alternate hosts in case the alternate hosts play an important role in pest population build up will also be effective in delaying the development of resistance to Bt toxins. Therefore, appropriate agricultural practices should be followed to reduce the carryover of pests from one season to another. This can help prevent potentially resistant pests from generating a next generation of resistant pests. Destruction of carryover populations is also part of remedial action plan in the event that emergence of Bt resistance is detected in monitoring programs (see Matten and Reynolds, this volume, pp. 137-178).

Use of a Planting Window

Defining a planting window for crop sowing such that the most susceptible stage of the crop escapes pest damage at peak periods of insect abundance can also be useful in maximizing the benefits of transgenic crops and prolong the life of transgenic crops. Planting the crops with first good monsoon rains has been found to be effective in controlling the damage by sorghum shoot fly (*Atherigona soccata*) and sorghum midge (*Stenodiplosis sorghicola*) (Sharma, 1993; Sharma and Ortiz, 2002). Similar strategies can be employed to avoid insect damage in other crops, and the reduced exposure of pests to Bt or other protective genes may prolong the effectiveness of transgenic crops. These practices should reduce the size of populations exposed to plant incorporated Bt, and thus reduce selection of resistant individuals.

Pesticide Application

Use of pesticide formulations such as soil application of granular systemic insecticides and spraying soft insecticides such as nuclear polyhedrosis virus (NPV) or endosulfan may be considered to suppress populations in the beginning of the season while minimizing adverse effects on the natural enemies of the pests. Broad-spectrum and toxic insecticides should be used only during the peak activity periods of the target pest (Forrester, 1990). These treatments should eliminate any Bt resistant insects as readily as Bt susceptible individuals. Efforts should be made to rotate pesticides with different mode of action, and avoid repetition of insecticides belonging to the same group or insecticides that fail to give effective control of the target pest, e.g., legume pod borer, *H. armigera* (Fitt, 1989; Forrester, 1990; Sharma, 2001).

Integrated Pest Management

Transgenic crops are compatible with other methods of pest control. Transgenic cotton (NuCotn 33) acts additively in combination with polyhedrosis virus AcNPV-Aalt in reducing the bollworm damage (All and Treacy, 1997). Insects such as *H. virescens, H. zea, Trichoplusia ni* and *Spodoptera exigua* are many times more sensitive to Karate sprays when they have a prior exposure to *Bacillus thuringiensis* (Harris et al., 1998). The enhanced insecticidal activity enables a more practical resistance management strategy for transgenic crops. Thus, transgenic crop can be used in conjunction with other methods of pest control without any detrimental or antagonistic effects.

Many pest control options that are disrupted by broad-spectrum pesticides are able to function with transgenic pesticide containing crops (Jasinski et al., 2001). Natural enemies of pests such as parasitoid insects (wasps, lacewings, etc.), predators (e.g., birds), and entomopathogenic nematodes and fungi typically fare far better in the presence of transgenic insect protected crops than the broad-spectrum synthetic pesticides these crops are displacing (Sharma and Ortiz, 2000). Other pest control treatments, such as bio-pesticides and natural plant products are fully compatible with transgenic crops as well.

LIMITATIONS AND RISKS ASSOCIATED WITH DEPLOYMENT OF INSECT-PROTECTED TRANSGENIC CROPS

The developments in plant biotechnology have both promise and problems. As a result of large-scale cultivation of transgenic plants with resistance to insect pests, secondary insect pests may become serious in the absence of sprays for the major insect pests (Hilder and Boulter, 1999; Sharma et al., 2000). If chemical insecticides are applied for the control of secondary insect pests, the potential advantage of transgenics would be diminished.

Insect exposure to transgenic plants over a long period of time may lead to development of resistance to toxin genes and thus may limit the usefulness of transgenics (Tabashnik, 1994; Gould, 1994, 1998). The evidence on these issues is still inconclusive, and there is a need for careful monitoring before the transgenic crops are deployed on a large commercial scale and under subsistence farming conditions.

TRANSGENICS IN PEST MANAGEMENT

The first transgenic crop was commercially grown in 1994, and large-scale cultivation was taken up in 1996 in the USA (McLaren, 1998). The area planted to transgenic crops increased from 1.7 million ha in 1996 to >50 million ha in 2000 (Federici, 1998; Griffiths, 1998).

Insect-resistant varieties in combination with the biocontrol agents have the potential to reduce dependence on insecticides and farmers' crop protection costs, thus benefiting farm livelihoods, the environment, and public health. The benefits to growers have been higher yields (Bullock and Nitsi, 2001), lower costs, and ease of management; an illustration of this phenomenon in cotton cultivation is given by the work of de Bianconi (this volume, pp. 223-235). In addition to the reduction in losses due to insect pests, the development and deployment of transgenic plants with insecticidal genes will also lead to: (i) reduction in insecticide expenses, (ii) reduced exposure of farm labor to insecticides, (iii) reduction in harmful effects of insecticides to non-target organisms, (iv) increased abundance of natural enemies of pests, (v) reduced amounts of insecticide residues in food and food products, and (vi) less pollution. Various advantages of transgenics in pest management are discussed below.

The effects of transgenic crops on insect population dynamics would be similar to the plants derived from conventional breeding for resistance to insect pests (Luginbill and Knipling, 1969). The effects of transgenic plants are anticipated to be continuous and cumulative over time, and result in lowering the insect populations to below the economic threshold level (ETL) (Sharma, 1993; Sharma and Ortiz, 2002), with minimal use of insecticides. The ETL is defined as the point at which the pest pressure has dropped to where control measures against the pest no longer give a savings over the damage that the pest will cause if unchecked. There is a need to understand the field performance of insect-resistant transgenic cultivars under diverse environmental conditions, long-term effects of the resistant cultivars on insect populations, level of adoption of the insect-resistant cultivars in space and time, and finally, their effect on economic thresholds (Gould, 1998; Hilder and Boulter, 1999).

Transgenic cultivars have shown considerable promise for minimizing insect pest damage alone and in combination with insecticides (Nelson and Pinto, 2001). One of the major contributions of transgenics in pest management is the reduction in the number of insecticide sprays, and the amount of insecticides needed for effective insect pest management (Schell, 1997; Lynch et al., 1999; Brickle et al., 1999). Production of insect-protected transgenic crops has resulted in reduced application of insecticides. Reduction in insecticide sprays will not only reduce the farmers' crop production costs, but also reduce the direct and indirect toxicity hazards to natural enemies, pollinators, and other non-target organisms (Jasinski et al., 2001).

CONCLUSIONS

Incorporation of insecticidal genes in crop plants offers tremendous improvements in pest management. Emphasis needs to be placed on combining exotic genes with conventional host plant resistance, and also with traits conferring resistance to other insect pests and diseases of regional importance to crop productivity. While several crops with commercial viability have been transformed in the developed world, very little has been done to use this technology to increase food production in the harsh environments of the tropics. Equally important is the need to design and adhere to rational biosafety regulations and to make this technology available to subsistence farmers, who cannot afford the high cost of seeds and chemical pesticides. The transgenic technology has tremendous potential, and every effort should be made to tap it for sustainable crop production under subsistence farming conditions in the tropics.

REFERENCES

Alam, M.F., Datta, K., Abrigo, E., Oliva, N., Tu, J., Virmani, S.S., and Datta, S.K. (1999) Transgenic insect-resistant maintainer line (IR68899B) for improvement of hybrid rice. *Plant Cell Reports* 18: 572-575.

All, J.N., and Treacy, M.F. (1997) Improved control of *Heliothis virescens* and *Helicoverpa zea* with a recombinant form of *Autographa californica* nuclear polyhedrosis virus and interaction with Bollgard R cotton. In: *Proceedings, Beltwide Cotton Conference*, 6-10 Jan 1997, New Orleans, USA. Volume 2. Memphis, USA: National Cotton Council. pp. 1294-1296.

Archer, T.L., Schuster, G., Patrick, C., Cronholm, G., Bynum, E.D. Jr., and Morrison, W.P. (2000) Whorl and stalk damage by European and Southwestern corn borers to four events of *Bacillus thuringiensis* transgenic maize. *Crop Protection* 19: 181-190.

Arencibia, A., Vazquez, R.I., Prieto, D., Tellez, P., Carmona, E.R., Coego, A., Hernandez, L., Riva, G.A. de-la, and Selman-Housein, G. (1997) Transgenic sugarcane plants resistant to stem borer attack. *Molecular Breeding* 3: 247-255.

Armstrong, C.L., Parker, G.B., Pershing, J.C., Brown, S.M., Sanders, P.R., Duncan, D.R., Stone, T., Dean, D.A., DeBoer, D.L., Hart, J., Howe, A.R., Morrish, F.M., Pajeau, M.E., Peterse, W.L., Reich, B.J., Rodriguez, R., Santino, C.G., Sato, S.J., Schuler, W., Sims, S.R., Stehling, S., Tarochione, L.J., and Fromm, M.E. (1995) Field evaluation of European corn borer control in progeny of 173 transgenic corn events expressing an insecticidal protein from *Bacillus thuringiensis*. *Crop Science* 35: 550-557.

Arpaia, S., Chiriatti, K., and Giorio, G. (1998) Predicting the adaptation of Colorado potato beetle (Coleoptera: Chrysomelidae) to transgenic eggplants expressing CryIII toxin: the role of gene dominance, migration, and fitness costs. *Journal of Economic Entomology* 91: 21-29.

Arpaia, S., De Marzo, L., Di Leo, G.M., Santoro, M.E., Mennella, G., and Vanloon, J.J.A. (2000) Feeding behaviour and reproductive biology of Colorado potato beetle adults fed transgenic potatoes expressing the *Bacillus thuringiensis* Cry3B endotoxin. *Entomologia Experimentalis et Applicata* 95: 31-37.

Arpaia, S., Mennella, G., Onofaro, V., Perri, E., Sunseri, F., and Rotino, G.L. (1997) Production of transgenic eggplant (*Solanum melongena* L.) resistant to Colorado potato beetle (*Leptinotarsa decemlineata* Say). *Theoretical and Applied Genetics* 95: 329-334.

Barton, K., Whiteley, H., and Yang, N.S. (1987) *Bacillus thuringiensis* δ-endotoxin in transgenic *Nicotiana tabacum* provides resistance to lepidopteran insects. *Plant Physiology* 85: 1103-1109.

Benedict, J.H., Sachs, E.S., Altman, D.W., Deaton, D.R., Kohel, R.J., Ring, D.R., and Berberich, B.A. (1996) Field performance of cotton expressing Cry IA insecticidal crystal protein for resistance to *Heliothis virescens* and *Helicoverpa zea* (Lepidoptera: Noctiudae). *Journal of Economic Entomology* 89: 230-238.

Bergvinson, D., Willcox, M.N., and Hoisington, D. (1997) Efficacy and deployment of transgenic plants for stem borer management. *Insect Science and its Application* 17: 157-167.

Berliner, E. (1915) Uber die Schalffsuchi der Mehlmottenraupo (*Ephestia kuhniella* Zell.) und thren Erreger, *Bacillus thuringiensis* n. sp. *Zietschrift fur Angewndte Entomologie* 2: 29-56.

Bohorova, N., Zhang, W., Julstrum, P., McLean, S., Luna, B., Brito, R.M., Diaz, L., Ramos, M.E., Estanol, P., and Pacheco, M. (1999) Production of transgenic tropical maize with cryIAb and cryIAc genes via microprojectile bombardment of immature embryos. *Theoretical and Applied Genetics* 99: 437-444.

Brewer, G.J. (1991) Resistance to *Bacillus thuringiensis* subsp. *kurstaki* in the sunflower moth (Lepidoptera: Pyralidae). *Environmental Entomology* 20: 316-322.

Brickle, D.S., Turnipseed, S.G., Sullivan, M.J., and Dugger, P. (1999) The efficacy of different insecticides and rates against bollworms (Lepidoptera: Noctuidae) in B.T. and conventional cotton. In: *Proceedings, Beltwide Cotton Production and Research Conference*, 3-7 January, 1999, Orlando, Florida, USA (Richter, D., ed.). Volume 2. Memphis, USA: National Cotton Council. pp. 934-936.

Bullock, D., and Nitsi, E.I. (2001) GMO adoption and private cost savings: GR soybeans and *Bt* corn. In: *Genetically Modified Organisms in Agriculture: Economics and Politics* (Nelson, G.C., ed.). New York, USA: Academic Publishers. pp. 21-38.

Canedo, V., Benavides, J., Golmirzaie, A., Cisneros, F., Ghislain, M., and Lagnaoui, A. (1999) Assessing Bt-transformed potatoes for potato tuber moth, *Phthorimaea operculella* (Zeller), management. In: *Impact on a Changing World. International Potato Center Program Report 1997-1998.* Lima, Peru: International Potato Center (CIP). pp. 161-169.

Cao, J., Tang, J.D., Strizhov, N., Shelton, A.M., and Earle, E.D. (1999) Transgenic broccoli with high levels of *Bacillus thuringiensis* Cry1C protein control diamondback moth larvae resistant to Cry1A or Cry1C. *Molecular Breeding* 5: 131-141.

Chakrabarti, S.K., Mandaokar, A.D., Kumar, P.A., and Sharma, R.P. (1998) Synergistic effect of Cry1A(c) and Cry1F delta-endotoxins of *Bacillus thuringiensis* on cotton bollworm, *Helicoverpa armigera*. *Current Science* 75: 663-664.

Chen, L., Marmey, P., Taylor, N.J., Brizard, J., Espinoza, C., D'Cruz, P., Huet, H., Zhang, S., de Kocho, A., Beachy, R.N., and Fauquet, C.M. (1998) Expression and inheritance of multiple transgenes in rice plants. *Nature Biotechnology* 16: 1060-1064.

Clark, T.L., Foster, J.E., Kamble, S.T., and Heinrichs, E.A. (2000) Comparison of Bt (*Bacillus thuringiensis* Berliner) maize and conventional measures for control of the European corn borer (Lepidoptera: Crambidae). *Journal of Entomological Science* 35: 118-128.

Cornu, D., Leple, J.C., Bonade-Bottino, M., Ross, A., Augustin, S., Delplanque, A., Jouanin, L., Pilate, G. and Ahuja, M.R. (1996) Expression of a proteinase inhibitor and a *Bacillus thuringiensis* delta-endotoxin in transgenic poplars. In: *Somatic Cell Genetics and Molecular Genetics of Trees* (Boerjan, W., and Neale, D.B., eds.). Dordrecht, Netherlands: Kluwer Academic Publishers. pp. 131-136.

Crickmore, N., Ziegler, D.R., Fietelson, J., Schnepf, E., Van Rie, J., Lereclus, D., Baum, J., and Dean, D.H. (1998) Revision of the nomenclature for *Bacillus thuringiensis* pesticidal crystal proteins. *Microbiology and Molecular Biology Review* 62: 807-813.

Datta, K., Vasquez, A., Tu, J., Torrizo, L., Alam, M.F., Oliva, N., Abrigo, E., Khush, G.S., and Datta, S.K. (1998) Constitutive and tissue-specific differential expression of the cryIA(b) gene in transgenic rice plants conferring resistance to rice insect pests. *Theoretical and Applied Genetics* 97: 20-30.

Dayal, S., Lavanya, M., Devi, P. and Sharma, K.K. (2003) An efficient protocol for shoot regeneration and genetic transformation of pigeonpea [*Cajanus cajan* (L.) Millsp.] by using leaf explants. *Plant Cell Reports* 21: 1072-1079.

Delannay, X., LaVallee, B.J., Proksch, R.K., Fuchs, R.L., Sims, S.K., Greenplate, J.T., Marrone, P.G., Dodson, R.B., Augustine, J.J., Layton, J.G., and Fischhoff, D.A. (1989) Field performance of transgenic tomato plants expressing *Bacillus thuringiensis* var *kurstaki* insect control protein. *Bio/Technology* 7: 1265-1269.

Diao, Y.P., and Xie, F.L. (1997) Bollworm resistant gene Bt examined and appraised by experts. *China Cottons* 24(8): 27.

DuRant, J.A., Roof, M.E., May, O.L., and Anderson, J.P. (1996) Influence of refugia on movement and distribution of bollworm/tobacco budworm larvae in bollgard cotton. In: *Proceedings, Beltwide Cotton Conference*, 9-12 January 1996, Nashville, Tennessee, USA. Volume 2. Memphis, USA: National Cotton Council. pp. 921-923.

Federici, B.A. (1998) Broad-scale leaf pest-killing plants to be true test. *California Agriculture* 52: 14-20.

Fischhoff, D.A., Bowdish, K.S., Perlak, F.J., Marrone, P.G., McCormick, S.M., Niedermeyer, J.G., Dean, D.A., Kusano-Kretzmer, K., Mayer, E.J., Rochester, D.E., Rogers, S.G., and Fraley, R.T. (1987) Insect tolerant tomato plants. *BioTechnology* 5: 807-812.

Fitt, G.P. (1989) The ecology of *Heliothis* species in relation to agroecosystems. *Annual Review of Entomology* 34: 17-52.

Fitt, G.P., and Forester, N.W. (1988) Overwintering of *Heliothis*–the importance of stubble cultivation. *Australian Cotton Grower* 8: 7-8.

Fitt, G., Mares, C.L., and Llewellyn, D.J. (1994) Field evaluation and potential ecological impact of transgenic cottons (*Gossypium hirsutum*) in Australia. *Biocontrol Science and Technology* 4: 535-548.

Forrester, N.W. (1990) Designing, implementing and servicing an insecticide resistance management strategy. *Pesticide Science* 28: 167-179.

Fujimoto, H., Itoh, K., Yamamoto, M., Kayozuka, J., and Shimamoto, K. (1993) Insect resistant rice generated by a modified delta endotoxin genes of *Bacillus thuringiensis*. *Bio/Technology* 11: 1151-1155.

Ghareyazie, B., Alinia, F., Menguito, C.A., Rubia, L.G., Palma, J.M. de., Liwanag, E.A.. Cohen, M.B., Khush, G.S., and Bennett, J. (1997) Enhanced resistance to two stem borers in an aromatic rice containing a synthetic CryIA(b) gene. *Molecular Breeding* 3: 401-414.

Gibson, D.M., Gallo, L.G., Krasnoff, S.B., and Ketchum, R.E.B. (1995) Increased efficiency of *Bacillus thuringiensis* subsp. *kurstaki* in combination with tannic acid. *Journal of Economic Entomology* 88: 270-277.

Gill, S.S., Cowles, E.A., and Pietrantonio, F.V. (1992) The mode of action of *Bacillus thuringiensis* endotoxins. *Annual Review of Entomology* 37: 615-636.

Gould, F. (1994) Potential and problems with high-dose strategies for pesticidal engineered crops. *Biocontrol Science and Technology* 4: 451-461.

Gould, F. (1998) Sustainability of transgenic insecticidal cultivars: integrating pest genetics and ecology. *Annual Review of Entomology* 43: 701-726.

Griffiths, W. (1998) Will genetically modified crops replace agrochemicals in modern agriculture? *Pesticide Outlook* 9: 6-8.

Guo, S.D., Cui, H.Z., Xia, L.Q., Wu, D.L., Ni, W.C., Zhang, Z.L., Zhang, B.L., and Xu, Y.J. (1999) Development of bivalent insect-resistant transgenic cotton plants. *Scientia Agricultura Sinica* 32: 1-7.

Hadi, M.Z., McMullen, M.D., and Finer, J.J. (1996) Transformation of 12 different plasmids into soybean via particle bombardment. *Plant Cell Reports* 15: 500-505.

Halcomb, J.L., Benedict, J.H., Correa, J.C., and Ring, D.R. (1996) Inter-plant movement and suppression of tobacco budworm in mixtures of transgenic Bt and non-transgenic cotton. In: *Proceedings, Beltwide Cotton Conference*, 9-12 January 1996, Nashville, TN, USA. Volume 2. Memphis, USA: National Cotton Council. pp. 924-927.

Harris, J.G., Hershey, C.N., Watkins, M.J., and Dugger, P. (1998) The usage of Karate (lambda-cyhalothrin) oversprays in combination with refugia, as a viable and sustainable resistance management strategy for B.T. cotton. In: *Proceedings, Beltwide Cotton Conference*, 5-9 January 1998, San Diego, California, USA. Volume 2. Memphis, USA: National Cotton Council. pp. 1217-1220.

Harshavardhan, D., Rani, T.S., Sharma, H.C., Richa, A., and Seetharama, N. (2002) Development and testing of *Bt* transgenic sorghum. In: *International Symposium on Molecular Approaches to Improve Crop Productivity and Quality*, 22-24 May 2002. Tamil Nadu Agricultural University, Coimbatore, Tamil Nadu, India.

Hilder, V.A., and Boulter, D. (1999) Genetic engineering of crop plants for insect resistance-a critical review. *Crop Protection* 18: 177-191.

Hofte, H., and Whiteley, H.R. (1989) Insecticidal crystal proteins of *Bacillus thuringiensis. Microbiology Reviews* 53: 242-255.

Hoy, C.W. (1999) Colorado potato beetle resistance management strategies for transgenic potatoes. *American Journal of Potato Research* 76: 215-219.

Huang, J., S. Rozelle, C. Pray, and Wang Q. (2002) Plant biotechnology in China. *Science* 295: 674-676.

ICRISAT (International Crops Research Institute for the Semi-Arid Tropics). (1992) *The Medium Term Plan*, Volume 1. Patancheru, Andhra Pradesh, India: International Crops Research Institute for the Semi-Arid tropics (ICRISAT). 80 pp.

Jansens, S., Cornelissen, M., Clercq, R. de, Reynaerts, A., and Peferoen, M. (1995) *Phthorimaea opercullella* (Lepidoptera: Gelechiidae) resistance in potato by expression of *Bacillus thuringiensis* Cry IA(b) insecticidal crystal protein. *Journal of Economic Entomology* 88: 1469-1476.

Jansens, S., Vliet, A. van, Dickburt, C., Buysse, L., Piens, C., Saey, B., Wulf, A. de, Gossele, V., Paez, A., and Gobel, E. (1997) Transgenic corn expressing a Cry9C insecticidal protein from *Bacillus thuringiensis* protected from European corn borer damage. *Crop Science* 37: 1616-1624.

Jasinski, J., Eisley, B., Young, C., Willson, H., and Kovach, J. (2001) Beneficial arthropod survey in transgenic and non-transgenic field crops in Ohio. *Special Circular, Ohio Agricultural Research and Development Center* 179: 99-102.

Jayanand, B., Sudarsanam, G., and Sharma, K.K. (2003) An efficient protocol for regeneration of whole plant of chickpea (*Cicer arietinum* L.) by using axillary meristem explants derived from *in vitro* germinated seedlings. *In Vitro Cellular and Developmental Biology-Plants* 39: 171-179.

Kar, S., Basu, D., Das, S., Ramkrishnan, N.A., Mukherjee, P., Nayak, P., and Sen, S.K. (1997) Expression of CryIA(c) gene of *Bacillus thuringiensis* in transgenic chickpea plants inhibits development of podborer (*Heliothis armigera*) larvae. *Transgenic Research* 6: 177-185.

Kennedy, G.G., and Whalon, M.E. (1995) Managing pest resistance to *Bacillus thuringiensis* endotoxins: constraints and incentives to implementation. *Journal of Economic Entomology* 88: 454-460.

Kota, M., Daniell, H., Varma, S., Garczynski, S.F., Gould, F., and Moar, W.J. (1999) Over expression of the *Bacillus thuringiensis* (Bt) Cry2Aa2 protein in chloroplasts confers resistance to plants against susceptible and Bt-resistant insects. *Proceedings, National Academy of Sciences, USA* 96: 1840-1845.

Koziel, M.G., Beland, G.L., Bowman, C., Carozzi, N.B., Crenshaw, R., Crossland, L., Dawson, J., Desai, N., Hill, M., Kadwell, S., Launis, K., Lewis, K., Maddox, D., McPherson, K., Meghji, M.R., Merlin, E., Rhodes, R., Warren, G.W., Wright, M., and Evola, S.V. (1993) Field performance of elite transgenic maize plants expressing an insecticidal protein derived from *Bacillus thuringiensis. BioTechnology* 11: 194-200.

Kumar, P.A., Mandaokar, A., Sreenivasu, K., Chakrabarti, S.K., Bisaria, S., Sharma, S.R., Kaur, S., and Sharma, R.P. (1998) Insect-resistant transgenic brinjal plants. *Molecular Breeding* 4: 33-37.

Lambert, A.L., Bradley, J.R. Jr., and Duyn, J.W. van. (1996) Effects of natural enemy conservation and planting date on the susceptibility of Bt cotton to *Helicoverpa zea* in North Carolina. In: *Proceedings, Beltwide Cotton Conference,* 9-12 January 1996, Nashville, Tennesse USA. Volume 2. Memphis, USA: National Cotton Council. pp. 931-935.

Lauer, J., and Wedberg, J. (1999) Grain yield of initial Bt corn hybrid introductions to farmers in the Northern Corn Belt. *Journal of Production Agriculture* 12: 373-376.

Li, W.B., Zarka, K.A., Douches, D.S., Coombs, J.J., Pett, W.L., Grafius, E.J., and Li, W.B. (1999) Co-expression of potato PVYo coat protein and cryV-Bt genes in potato. *Journal of the American Society for Horticultural Science* 124: 218-223.

Luginbill, P. Jr., and Knipling, E.F. (1969) Suppression of wheat stem fly with resistant wheat. United States Department of Agriculture/Agriculture Research Service (USDA/ARS). *Production Research Reports* 107: 1-9.

Lynch, R.E., Wiseman, B.R., Plaisted, D., and Warnick, D. (1999) Evaluation of transgenic sweet corn hybrids expressing CryIA(b) toxin for resistance to corn earworm and fall armyworm (Lepidoptera: Noctuidae). *Journal of Economic Entomology* 92: 246-252.

MacIntosh, S.C., Kishore, G.M., Perlak, F.J., Marrone, P.G., Stone, T.B., Sims, S.R., and Fuchs, R.L. (1990) Potentiation of *Bacillus thuringiensis* insecticidal activity by serine protease inhibitors. *Journal of Agriculture and Food Chemistry* 38: 1145-1152.

Mandaokar, A.D., Goyal, R.K., Shukla, A., Bisaria, S., Bhalla, R., Reddy, V.S., Chaurasia, A., Sharma, R.P., Altosaar, I., and Kumar, P.A. (2000) Transgenic tomato plants resistant to fruit borer (*Helicoverpa armigera* Hubner). *Crop Protection* 19: 307-312.

McGaughey, W.H., and Whalon, M.E. (1992) Managing insect resistance to *Bacillus thuringiensis* toxins. *Science* 258: 1451-1455.

McLaren, J.S. (1998) The success of transgenic crops in the USA. *Pesticide Outlook* 9: 36-41.

Mohammed, A., Douches, D.S., Pett, W., Grafius, E., Coombs, J., Liswidowati, W.L., and Madkour, M.A. (2000) Evaluation of potato tuber moth (Lepidoptera: Gelechiidae) resistance in tubers of Bt-Cry5 transgenic potato lines. *Journal of Economic Entomology* 93: 472-476.

Mqbool, S.B., Husnain, T., Raizuddin, S., and Christou, P. (1998) Effective control of yellow rice stem borer and rice leaf folder in transgenic rice *indica* varieties Basmati 370 and M 7 using novel δ-endotoxin Cry2A *Bacillus thuringiensis* gene. *Molecular Breeding* 4: 501-507.

Munkvold, G.P., Hellmich, R.L., and Rice, L.G. (1999) Comparison of fumonisin concentrations in kernels of trangenic Bt maize hybrids and nontransgenic hybrids. *Plant Disease* 83: 130-138.

Murray, D.A.H., and Zalucki, M.P. (1990) Effect of soil moisture and simulated rainfall on pupal survival and moth emergence of *Helicoverpa punctigera* (Wallengen) and *H. armigera* (Hubner) (Lepidoptera: Noctuidae). *Journal of the Australian Entomological Society* 29: 193-197.

Nayak, P., Basu, D., Das, S., Basu, A., Ghosh, D., Ramakrishnan, N.A., Ghosh, M., and Sen, S.K. (1997) Transgenic elite *indica* rice plants expressing CryIAc deltaendotoxin of *Bacillus thuringiensis* are resistant against yellow stem borer (*Scirpophaga incertulas*). *Proceedings, National Academy of Sciences, USA* 94: 2111-2116.

Nelson, G.C. (2001) Traits and techniques of GMOs. In: *Genetically Modified Organisms in Agriculture: Economics and Politics* (Nelson, G.C., ed.). New York, USA: Academic Publishers. pp. 7-19.

Nelson, G.C., and Pinto de, A. (2001) GMO adoption and market effects In: *Genetically Modified Organisms in Agriculture: Economics and Politics* (Nelson, G.C., ed.). New York, USA: Academic Publishers. pp. 57-59.

Oerke, E.C., Dehne, H.W., Schonbeck, F., and Weber, A. (1994) *Crop Production and Crop Protection: Estimated Losses in Major Food and Cash Crops.* Amsterdam, The Netherlands: Elsevier Science Publishers.

Onstad, D.W., and Gould, F. (1998) Modeling the dynamics of adaptation to transgenic maize by European corn borer (Lepidoptera: Pyralidae). *Journal of Economic Entomology* 91: 585-593.

Perlak, F.J., Stone, T.B., Muskopf, Y.N., Petersen, L.J., Parker, G.B., McPherson, S.A., Wyman, J., Love, S., Reed, G., Biever, D., and Fischhoff, D.A. (1993) Genetically improved potatoes: protection from damage by Colorado potato beetles. *Plant Molecular Biology* 22: 313-321.

Prasanna, L., Moss, J.P., and Sharma, K.K. (1997) In vitro culture provides additional variation for pigeonpea crop improvement. *In Vitro Cellular and Developmental Biology–Plants* 33: 30-37.

Qaim, M., and D. Zilberman (2003) Yield effects of genetically modified crops in developing countries. *Science* 299: 900-902.

Rensburg van, J.B.J. (1999) Evaluation of Bt-transgenic maize for resistance to the stem borers *Busseola fusca* (Fuller) and *Chilo partellus* (Swinhoe) in South Africa. *South African Journal of Plant and Soil* 16: 38-43.

Sachs, E.S., Benedict, J.H., Taylor, J.F., Stelly, D.M., Davis, S.K., and Altman, D.W. (1996) Pyramiding CryIA(b) insecticidal protein and terpenoids in cotton to resist tobacco budworm (Lepidoptera: Noctuidae). *Environmental Entomology* 25: 1257-1266.

Schell, J. (1997) Cotton carrying the recombinant insect poison Bt toxin: no case to doubt the benefits of plant biotechnology. *Current Opinion in Biotechnology* 8: 235-236.

Schnepf, H.E., and Whiteley, H.R. (1981) Cloning and expression of *Bacillus thuringnensis* crystal protein gene in *Escherichia coli. Proceedings, National Academy of Sciences, USA* 78: 2893-2897.

Schuler, T.H., Poppy, G.M., Kerry, B.R., and Donholm, L. (1998) Insect resistant transgenic plants. *Trends in Biotechnology* 16: 168-175.

Selvapandian, A., Reddy, V.S., Ananda Kumar, P., Tiwari, K.K., and Bhatnagar, R.K. (1998) Transformation of *Nicotiana tobaccum* with a native cry11a5 gene confers complete protection against *Heliothis armigera. Molecular Breeding* 4: 473-478.

Sharma, H.C. (1993) Host plant resistance to insects in sorghum and its role in integrated pest management. *Crop Protection* 12: 11-34.

Sharma, H.C. (2001) Cotton bollworm/legume pod borer, *Helicoverpa armigera* (Hubner) (Noctuidae: Lepidoptera): Biology and management. *Crop Protection Compendium.* Wallingford, UK: Commonwealth Agricultural Bureaux.

Sharma, H.C., Ananda Kumar, P., Seetharama, N., Hari Prasad, K.V., and Singh, B.U. (1999) Role of transgenic plants in pest management in sorghum. In: *Symposium on Tissue Culture and Genetic Transformation of Sorghum, 23-28 Feb 1999.* International Crops Research Institute for the Semi-Arid Tropics (ICRISAT), Patancheru 502 324, Andhra Pradesh, India.

Sharma, H.C., and Ortiz, R. (2000) Transgenics, pest management, and the environment. *Current Science* 79: 421-437.

Sharma, H.C., and Ortiz, R. (2002) Host plant resistance to insects: An eco-friendly approach for pest management and environment conservation. *Journal of Environmental Biology* 23: 111-135.

Sharma, H.C., Sharma, K.K, Seetharama, N., and Ortiz, R. (2001) Genetic transformation of crop plants: Risks and opportunities for the rural poor. *Current Science* 80: 1495-1508.

Sharma, H.C., Sharma, K.K., Seetharama, N., and Ortiz, R. (2000) Prospects for transgenic resistance to insects. *Electronic Journal of Biotechnology* 3(2): 25. <http://www.ejbiotechnology.info/content/vol3/issue2/full/3/index.html>.

Sharma, H.C., Crouch, J.H., Sharma, K.K, Seetharama, N., and Hash, C.T. (2002) Applications of biotechnology for crop improvement: prospects and constraints. *Plant Science* 163: 381-395.

Sharma, K.K., and Ortiz, R. (2000) Program for the application of genetic transformation for crop improvement in the semi-arid tropics. *In Vitro Cellular and Developmental Biology–Plants* 36: 83-92

Sharma, K.K., Sharma, H.C., Seetharama, N., and Ortiz, R. (2002) Development and deployment of transgenic plants: Biosafety considerations. *In Vitro Cellular and Developmental Biology–Plants* 38: 106-115.

Shukla, S., Sharma, H.C., Sharma, K.K., and Shrivastava, S. (2003) Effect of transgenic tobacco plants with *Bacillus thuringiensis* Cry1A(b) and soybean trypsin inhibitor genes on consumption and utilization of food, and biology of *Helicoverpa armigera* (Noctuidae: Lepidoptera). *Plant Science* (submitted).

Singsit, C., Adang, M.J., Lynch, R.E., Anderson, W.F., Aiming Wang, Cardineau, G., and Ozias-Akins, P. (1997) Expression of a *Bacillus thuringiensis* CryIA(c) gene in transgenic peanut plants and its efficacy against lesser cornstalk borer. *Transgenic Research* 6: 169-176.

Stoger, E., Williams, S., Christou, P., Down, R.E., and Gatehouse, J.A. (1999) Expression of the insecticidal lectin from snowdrop (*Galanthus nivalis* agglutinin; GNA) in transgenic wheat plants: effects on predation by the grain aphid *Sitobion avenae*. *Molecular Breeding* 5: 65-73.

Tabashnik, B.E. (1994) Evolution of resistance to *Bacillus thuringiensis*. *Annual Review of Entomology* 39: 47-79.

Tailor, R., Tippett, J., Gibb, G., Pells, S., Pike, D., Jordan, L., and Ely, S. (1992) Identification and characterization of a novel *Bacillus thuringiensis*-endotoxin entomocidal to coleopteran and lepidopteran larvae. *Molecular Microbiology* 7: 1211-1217.

Vaeck, M., Reynaerts, A., Hofte, H., Jansens, S., DeBeuckleer, M., Dean, C., Zabeau, M., Van Montagu, M., and Leemans, J. (1987) Transgenic plants protected from insect attack. *Nature* 327: 33-37.

Van der Salm, T., Bosch, D., Honee, G., Feng, I., Munsterman, E., Bakker, P., Stiekema, W.J., and Visser, B. (1994) Insect resistance of transgenic plants that express modified Cry1A(b) and Cry 1C genes: a resistance management strategy. *Plant Molecular Biology* 26: 51-59.

Westedt, A.L., Douches, D.S., Pett, W., and Grafius, E.J. (1998) Evaluation of natural and engineered resistance mechanisms in *Solanum tuberosum* for resistance to *Phthorimaea operculella* (Lepidoptera: Gelechiidae). *Journal of Economic Entomology* 91: 552-556.

Williams, W.P., Sagers, J.B., Hanten, J.A., Davis, F.M., and Buckley, P.M. (1997) Transgenic corn evaluated for resistance to fall armyworm and Southwestern corn borer. *Crop Science* 37: 957-962.

Wilson, W.D., Flint, H.M., Deaton, R.W., Fischhoff, D.A., Perlak, F.J., Armstrong, T.A., Fuchs, R.L., Berberich, S.A., Parks, N.J., and Stapp, B.R. (1992) Resistance of cotton lines containing a *Bacillus thuringiensis* toxin to pink bollworm (Lepidoptera: Gelechiidae) and other insects. *Journal of Economic Entomology* 85: 1516-1521.

Xiang, Y., Wong, W.K.R., Ma, M.C., and Wong, R.S.C. (2000) *Agrobacterium*-mediated transformation of *Brassica campestris* ssp. *parachinensis* with synthetic *Bacillus thuringiensis* Cry1A(b) and Cry1A(c) genes. *Plant Cell Reports* 19: 251-256.

Xu, D., McElroy, D., Thoraburg, R.W., and Wu, R. (1993) Systemic induction of a potato pin 2 promoter by wounding methyl jasmonate and abscisic acid in transgenic rice plants. *Plant Molecular Biology* 22: 573-588.

Yang, Y.T., Wang, D.H., Zhu, M.H., Yi, H.J. (1999) The key stage for control of *Helicoverpa armigera* Hubner by watering field. *Journal of Nanjing Agricultural University* 22: 108-110.

Zhao, J.Z., Shi, X.P., Fan, X.L., Zhang, C.Y., Zhao, R.M., and Fan, Y.L. (1998) Insecticidal activity of transgenic tobacco co-expressing Bt and CpTI genes on *Helicoverpa armigera* and its role in delaying the development of pest resistance. *Rice Biotechnology Quarterly* 34: 9-10.

Zhao, J.Z., Fan, X.L., Shi, X.P., Zhao R.M., and Fan, Y.L. (1997) Gene pyramiding: an effective strategy of resistance management for *Helicoverpa armigera* and *Bacillus thuringiensis*. *Resistant Pest Management* 9: 19-21.

Engineering of Bt Transgenic Rice for Insect Pest Protection

S. K. Datta
G. Chandel
J. Tu
N. Baisakh
K. Datta

SUMMARY. Stem borer is a serious pest that causes considerable yield losses of rice in Asia. Pesticides have been used broadly to control this pest. Transgenic Bt rice for a wide variety of cultivation conditions have been developed and reported on earlier. We review work done to develop Bt rice with multiple Bt genes that have different receptor binding domains, and with a fusion *cry* gene, *cry1B/cry1Ab*. Generation of Bt rice has been performed in different cultivars and recently in the hybrid

S. K. Datta, G. Chandel, N. Baisakh and K. Datta are affiliated with the International Rice Research Institute, Plant Breeding, Genetics, and Biochemistry Division, DAPO Box 7777, Metro Manila, Philippines (E-mail: s.datta@cgiar.org).

J. Tu is affiliated with the International Rice Research Institute, Plant Breeding, Genetics, and Biochemistry Division, DAPO Box 7777, Metro Manila, Philippines, and is also affiliated with The Chinese University of Hong Kong, Shatin, Hong Kong.

The authors thank CIBA-GEIGY (Syngenta), Prof. I. Altosaar, Prof. Y. Fan, and Dr. R. Frutos for providing the Bt genes, and Dr. M. Cohen and Ms. R. Aguda for helping in insect bioassay and valuable discussions. The authors also thank Drs. J. Tu, M. F. Alam, G. Ye, S. Balachandran, N. H. Ho, and K. M. Thet for their Bt research contributions in the lab. The authors acknowledge the technical assistance of M. Viray.

The financial support from Bundesministerium für Wirtschaftliche Zusammenarbeit und Entwicklung (BMZ, Germany) and the Rockefeller Foundation (USA) is gratefully acknowledged.

[Haworth co-indexing entry note]: "Engineering of Bt Transgenic Rice for Insect Pest Protection." Datta, S. K. et al. Co-published simultaneously in *Journal of New Seeds* (Food Products Press, an imprint of The Haworth Press, Inc.) Vol. 5, No. 2/3, 2003, pp. 77-91; and: *Bacillus thuringiensis: A Cornerstone of Modern Agriculture* (ed: Matthew Metz) Food Products Press, an imprint of The Haworth Press, Inc., 2003, pp. 77-91. Single or multiple copies of this article are available for a fee from The Haworth Document Delivery Service [1-800-HAWORTH, 9:00 a.m. - 5:00 p.m. (EST). E-mail address: docdelivery@haworthpress.com].

'Shanyou 63', which gains protection against four insect pests. The Bt-rice hybrid line is free of the selectable marker gene, *hph*, and has been successfully field evaluated in China for 4 years. Transgene pyramiding with Bt genes alone or in combination with other genes for plant protection has been developed in rice. Bt rice in Asia has the potential to provide plant protection while curbing use of pesticides and reducing yield losses from infestation. The deployment of insect protection genes using a gene pyramiding strategy is presented as a vital strategy for maximizing crop protection and countering the development of resistant pests. *[Article copies available for a fee from The Haworth Document Delivery Service: 1-800-HAWORTH. E-mail address: <docdelivery@haworthpress.com> Website: <http://www.HaworthPress.com> © 2003 by The Haworth Press, Inc. All rights reserved.]*

KEYWORDS. Transgenic, stem borer, Bt, rice, field evaluation, pyramiding, hybrid

INTRODUCTION

Available information shows a 10-35% loss in rice production in different Asian rice-growing countries and a 2-35% loss in rice productivity worldwide due to major insect pests (Ramaswamy and Jatileksono, 1996). Stem borer is the principal devastating insect pest of rice: striped stem borer (*Chilo suppressalis*) and yellow stem borer (*Scirlophage incertulas*). The rice leaffolder (*Cnaphalocrosis medinalis*) is also a significant pest. Stem borer larvae feed inside the stem pith cells and cause 'deadheart' during infestation of rice that is in the vegetative growth stage (that does not bear the panicle). Infestation of rice in its reproductive stages causes the panicles to become chaffy with a 'whitehead' appearance. Deployment of stem borer protected varieties is a potentially environment-friendly and cost-effective component of integrated pest management (IPM). However, conventional resistance breeding has been handicapped due to the unavailability of germplasm resistant to this pest. Genetic engineering has enabled development of many transgenic crops including rice carrying *cry* insecticidal crystal protein (ICP) genes from *Bacillus thuringiensis* (Bt), which confers protection against stem borer. Here we review the progress of transgenic Bt rice research at the International Rice Research Institute (IRRI), Philippines. Using all three established transformation methods–*Agrobacterium*-mediated, biolistic and protoplast (Datta et al., 1996, 1997; Datta, 2000a; Datta and Datta, 2002)–a large number of transgenic

Bt-rice varieties have been produced (Table 1). A number of other groups have also successfully developed and reported on Bt rice (Ghareyazie et al., 1997; Nayak et al., 1997; Cheng et al., 1998; Maqbool et al., 1998; Breitler et al., 2000; Marfà et al., 2002).

PROMOTERS USED WITH SINGLE AND FUSION Bt GENES

The expression of various synthetic and modified *crylAb* and *crylAc* genes (individually and as fusions) driven by constitutive (35S, Actin-1, and Ubiquitin) as well as tissue-specific (PEPC and pith-specific) promoters has been studied in detail (Figure 1; Datta et al., 1998). The germplasm used to develop Bt rice represent cultivars from different countries adapted to different climates including lowland, deepwater and irrigated conditions: IRRI-New Plant Type (NPT), maintainer and restorer lines for hybrid rice breeding, aromatic rice, and locally-adapted improved cultivars (Table 1).

Apart from use of individual *cry* genes, a fusion gene, *crylAb/crylAc* (Figure 1; Tu et al., 1998) has been used to transform indica rice IR72. Another construct with a translational fusion gene *crylB/crylAb* has been used to transform elite indica rice lines from Vietnam (in collaboration with Roger Frutos, CIRAD, Montpellier, France; Ho et al., 2003). All such constructs are being used as a strategy to improve pest protection, delay resistance development in pests, and widen targeted pest specificity. Immunoblot analysis and ELISA showed the variation of Cry protein levels from 0.01% to a maximum of 1% of the total soluble protein in different transgenic rice lines (Alam et al., 1998; Datta et al., 1998; Tu et al., 1998; Baisakh, 2000). The cut-stem and whole-plant bioassays, performed in greenhouse trials demonstrated 100% larval mortality of the yellow stem borer feeding on plants expressing Cry protein within this range. These plants showed no measurable phenotypic trade-off (e.g., yield decline).

GENERATION AND FIELD EVALUATION OF TRANSGENIC, INBRED AND HYBRID Bt RICE

Besides, inbred lines, our Bt research was extended to the hybrid rice program (Figure 2) by developing a homozygous transgenic Bt maintainer (e.g., IR68899B: Alam et al., 1999) and a restorer line (e.g., Minghui 63: Tu et al., 2000).

A collaborative IRRI-China research program has been performing

TABLE 1. Status of the Transgenic *Bt* Rice Research at the Tissue Culture and Genetic Engineering Laboratory, International Rice Research Institute, Philippines

Cultivar	*Bt* Gene	Promoter	Generation status	Transformation	Expression	References
T309	*cry1Ab/cry1Ac*	Actin	T_3	B	Stable expression	Wu et al., 1997
Indica/japonica	*Bt*	Several	Variable	B/P	Variable	Datta et al., 1996
IR 72	*cry1Ab/cry1Ac*	Actin	T_8	B	Tested in field condition	Datta et al., 1998
IR 72	*cry1Ab*	PEPC	T_2	B	Stable expression	Datta et al., 1998
IR 64	*cry1Ab*	35 S	T_6	B	Unstable expression	Datta et al., 1998
CB II	*cry1Ab*	35 S, PEPC, Pith	T_2	P	Stable expression	Datta et al., 1998
IRRI-NPT	*cry1Ab*	35 S, PEPC	T_3	B	Stable expression	Datta et al., 1998, 1999
Basmati 370	*cry1Ab*	35 S	T_2	B	Stable expression	
IR 51500-Ac-11-1	*cry1Ab*	35 S	T_2	B	Stable expression	Datta et al., 1998
IR 58	*cry1Ab*	35 S	T_2	B	Stable expression	Datta et al., 1998
Vaidehi	*cry1Ab* *cry1Ab/cry1Ac*	35 SActin	T_3 T_3	B	Stable expression Stable expression	Alam et al., 1999 Baisakh, 2000
IR 68899 B	*cry1Ab*	35 S	T_2	B	Stable expression	Alam et al., 1999
MH 63	*cry1Ab/cry1Ac*	Actin	T_8	B	Stable expression, Tested in field condition	Tu et al., 2000
Tulasi	*cry1Ab/cry1Ac*	Actin	T_2	B	Stable expression	Baisakh, 2000
BR827-35R	*cry1Ab/cry1Ac*	Actin	T_0	B		Balachandran et al., 2003
IR 68897 B	*cry1Ab/cry1Ac*	Actin	T_1	B		Balachandran et al., 2003

	Gene	Promoter	Generation	Method	Expression	Reference
Azucena	cry1Ab	PEPC	T_1	B	Stable expression	Unpublished
	cry1Ab		T_1	B	Stable expression	
Pusa Bas.	cry1Ab/cry1Ac	Actin	T_1	B	Stable expression	Unpublished
BPT -5204	cry1Ab/cry1Ac	Actin	T_1	B	Stable expression	Unpublished
Milagrosa	cry1Ab/cry1Ac	Actin	T_0	B		Unpublished
KDML 105	cry1Ab	PEPC	T_0	B		Unpublished
Sakha 101	cry1Ab	35 S, PEPC, pith	T_0	B		Unpublished
Dinorado	cry1Ab	PEPC	T_1	B	Stable expression	Unpublished
Kunihikari	cry1Ab/cry1Ac	Actin	T_0	B		Unpublished
C4	cry1Ab	PEPC	T_0	B		
Mot Bui	cry1Ab/cry1B cry1Ab/cry1Ac	UbiquitinActin	T_1 T_1	B/Agro	Stable expression	Ho et al., 2003 (in preparation)
Nang Hong Cho Dao	cry1 Ab/cry1B	Ubiquitin	T_1	Agro	Stable expression	Ho et al., 2003 (in preparation)
IR72	cry1Ab/cry1Ac + Xa21 + PR genes	Actin 1 + 35 S	T_3	B	Stable expression	Datta et al., 2002

B = Biolistic, P = Protoplast, Agro = Agrobacterium-mediated transformation

FIGURE 1. Partial diagram of *Bt* gene constructs used in single or in combinations in development of *Bt* rice.

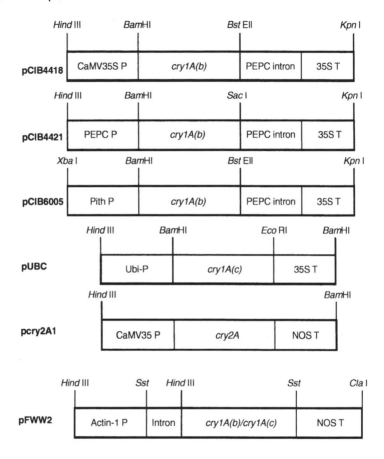

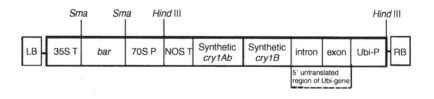

FIGURE 2. Schematic figure showing use of transgenic breeding approach for developing yellow stemborer (YSB)-resistant rice hybrids (Schemes I and II) (modified from Alam et al., 1998).

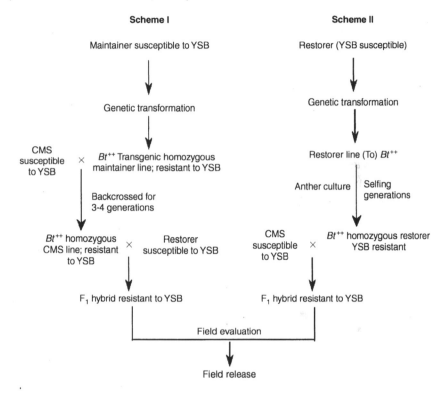

field evaluations of IRRI Bt hybrid rice lines for over 4 years (Datta, 2000b). The rice lines tested include IR72 (an IRRI-bred elite indica variety), MH 63 (a Chinese cytoplasmic male sterile restorer line), and a hybrid, Shayou 63 derived from Zhenshan 97A × MH 63. The field trial results clearly demonstrated the pest resistance of the transgenic lines, which showed very high protection against yellow stem borer, striped stem borer, and leaffolder under natural as well as artificial infestations (Tables 2 and 3). In the case of Bt hybrid Shanyou 63, it outperformed non-Bt Shanyou 63 by 28.9% in grain yield (Tu et al., 2000; Tables 4 and 5). These field trials, done in Wuhan, China, are the first ever of a commercial hybrid Bt rice.

The IRRI Bt hybrid was made free of the selectable marker gene *hph*, which was bred out of Bt-Minghui 63, the restorer parent of the hybrid

TABLE 2. Resistance Reactions of Minghui 63/*Bt* (TT51-1) and Shanyou 63 Against Natural and Manual Infestation of Yellow Stem Borer Under Field Conditions (Wuhan, China, 2000)*

Infestation	Line	No. of tillers/plant	No. of panicles/plant	Booting stage		Grain-filling stage	
				% plants w deadheart	% plants w whitehead	% plants w deadheart	% plants w whitehead
Natural	M.H. 63/*Bt*	20.8	18.3	3.3**	0.2 ± 0.1**	0.0**	0.0 ± 0.0**
	M.H. 63/ck	19.8	11.8	96.7	2.6 ± 0.4	93.3	36.4 ± 1.9
	S.Y. 63/*Bt*	17.5	15.4	10.0**	0.5 ± 0.3**	0.0**	0.0 ± 0.0**
	S.Y. 63/ck	17.9	10.6	83.3	4.4 ± 0.7	100.0	54.9 ± 2.1
Manual	M.H. 63/*Bt*	37.3	30.1	0.0**	0.0 ± 0.0**	13.3**	0.5 ± 1.1**
	M.H. 63/ck	48.8	16.5	100.0	15.2 ± 0.6	100.0	54.4 ± 2.9
	S.Y. 63/*Bt*	33.4	28.1	0.0**	0.0 ± 0.0**	13.3**	0.7 ± 1.0**
	S.Y. 63/ck	38.4	17.3	100.0	25.4 ± 0.5	100.0	70.8 ± 1.1

*Sixty to 70 plants tested per line per replication and about 150 larvae infested per plant in the case of manual infestation.
** Significantly different from control at $P < 0.01$.

TABLE 3. Resistance Reactions of Minghui 63/*Bt* (TT51-1) and Shanyou 63 Against Natural Infestations of Leaffolder Under Field Conditions (Wuhan, China, 2000)

Line	No. of plants tested	Insect test plot		Yield test plot	
		% plants affected	No. of damaged leaves/plant	% plants affected	No. of damaged leaves/plant
Minghui 63/*Bt*	20	3.3**	0.1 ± 0.1**	6.7**	0.1 ± 0.1**
Minghui 63/ck	20	100.0	19.8 ± 1.8	100.0	21.3 ± 2.5
Shanyou 63/*Bt*	20	6.7**	0.1 ± 0.0**	6.7**	0.1 ± 0.1**
Shanyou 63/ck	20	100.0	13.2 ± 2.3	100.0	20.8 ± 2.7

**Significantly different from control at $P < 0.01$.

TABLE 4. Yield Performance of Minghui 63/*Bt* (TT51-1) and Its Derived Hybrids Under a Field Treated Without Chemical Spray (Wuhan, China, 2000)

Line/hybrid	Plot yield* (kg)			Average plot yield (kg)	% above check (±%)
	I	II	III		
Minghui 63/*Bt*	6.20	5.84	6.56	6.20 ± 0.36	140.3**
Minghui 63/ck	2.58	2.98	2.18	2.58 ± 0.40	0.0
Shanyou 63/*Bt*	8.49	7.01	8.74	7.73 ± 0.74	200.8**
Shanyou 63/ck	2.33	2.16	3.22	2.57 ± 0.57	0.0
Eryou 63/*Bt*	8.29	7.54	6.90	7.58 ± 0.69	194.9**
Xieyou 63/*Bt*	8.29	6.99	7.24	7.51 ± 0.69	192.2**
Mayou 63/*Bt*	8.72	8.37	8.84	8.54 ± 0.24	249.4**

*Plot area is 6.7 m^2.
**Significantly different from control at $P < 0.01$.

TABLE 5. Yield Performance of Minghui 63/*Bt* (TT51-1) and Its Derived Hybrids Under a Field Treated with Chemical Spray (Wuhan, China, 2000)

Line/hybrid	Plot yield* (kg)			Average plot yield (kg)	% above check (±%)
	I	II	III		
Minghui 63/*Bt*	6.27	7.51	6.81	6.86 ± 0.62	21.2**
Minghui 63/ck	6.12	5.58	5.27	5.66 ± 0.43	0.0
Shanyou 63/*Bt*	6.01	7.74	7.35	7.03 ± 0.90	63.8**
Shanyou 63/ck	5.52	3.81	4.37	4.57 ± 0.87	0.0**
Eryou 63/*Bt*	7.62	6.53	8.57	7.57 ± 1.02	65.6**
Xieyou 63/*Bt*	7.30	7.43	7.56	7.43 ± 0.13	62.5**
Mayou 63/*Bt*	8.29	7.35	6.51	7.38 ± 0.81	61.5**

*Plot area is 6.7 m^2.
**Significantly different from control at $P < 0.01$.
Note: The yield increase of the *Bt* maintainer line or its derived hybrid over their nontransgenic control was not expected under chemical spray. But since the outbreaks of stem borers (yellow stem, leaffolder, and so on) were too heavy to control in the local area in 2000, the observed yields in both *Bt* maintainers and hybrids were eventually significantly higher than those in their controls.

(Datta, 2000b). This was possible due to independent segregation of the *hph* and Bt genes, integrated at two different chromosomal loci (Tu et al., 2002). This will ease the evaluation process in obtaining approval for commercializing Bt hybrid rice, since the presence of antibiotic resistance genes has been one of the many obstacles to acceptance of genetically engineered crops. Similarly, the transgenic Bt-IR72 was successfully field evaluated in Zhezhang University, China. The results indicated significantly higher protection of the transgenic lines against four lepidopteran insects, yellow, pink, and stripped stem borer, and leaffolder (Ye et al., 2001). IRRI-ICAR collaborative research program at Directorate of Rice Research (DRR), Hyderabad, India on field evaluation of Bt rice showed excellent protection of Bt-IR72 cultivar from yellow and striped stem borer after two seasons of successful testing under greenhouse conditions.

In addition to insect protection being an important measure of the utility of Bt rice, agronomic performance should also be maintained. There has been stipulation that transgene expression can reduce the productivity of a crop. However, our comparison of Bt and non-Bt Shanyou 63 under field conditions (Table 6) provides an example where the transgenic line is at least as productive as the unmodified line.

GENE PYRAMIDING FOR MANAGEMENT OF RESISTANCE IN TARGET PESTS

One proposed strategy for delaying or preventing the development of pest resistance to insecticide and insect repellant genes is to employ a multi-gene approach termed 'pyramiding' crops (Cohen et al., 1998; Frutos et al., 1999). A pyramiding approach that we have explored, with the hopes of overcoming challenges related to proper gene expression from multiple constructs, is to employ two Bt *cry* genes as a translational fusion. The design of Bt fusions was performed so that different receptor binding domains in the target pest would be used by the two components, a key requirement for pyramiding to reduce the chance for development of resistant pests. Management strategies developed for Bt crops (e.g., cotton or corn) commercialized in countries like the USA with industrial scale farming operations may not serve as a model for rice in Asian countries. For example, an intensive and highly coordinated insect resistance management (IRM) program for Bt cotton in the U.S. (see Matten and Reynolds, this volume, pp. 137-178) may be inapproachable in many areas of Asia. Millions of small holding farmers, diverse varieties and different climates present different challenges to the

TABLE 6. Agronomic Traits of the *Bt* and Non-*Bt* Hybrids Under Field Conditions (Wuhan, China, 1999)

Hybrid	Days to flower	Plant height (cm)	Panicles per plant	Filled grains per panicle	Non-filled grains per panicle	Total grains per panicle	Seed-setting rate (%)	1000 grain weight (g)	Expected yield (t/ha)	Observed yield* (t/ha)	+/- (%)
Bt Shanyou 63	95	108.6	9.98	140.0	21.9	161.7	87.6	28.00	7.44	8.69	+28.9**
Shanyou 63	95	106.4	9.60	143.7	28.8	172.5	83.3	28.76	7.44	6.74	----

* The observed yield was measured based on the average production per unit of four subplots after harvesting and then converted into tons per hectare.
** Significantly different from control at $P < 0.01$.

adoption and management of Bt rice in Asia. Any transgenic technology aimed at combating pests and countering resistance development must be developed in suitable backgrounds of different elite cultivars and traditional varieties, then field evaluated in multiple locations in Asia and other rice growing countries. In the areas where IRM parameters might be so variable, and the stakeholders so numerous as to make coordination, enforcement and monitoring impractical, gene 'pyramiding' as a strategy for countering insect resistance will be more pivotal than anywhere else.

PROMOTER EFFECTS ON DEVELOPMENTAL CONTROL OF Bt TRANSGENE EXPRESSION

It has been found that Bt rice with *cryIAb* driven by the phosphoenol-pyruvate carboxylase (PEPC) promoter showed green-tissue specific expression and was highly resistant to stem borers and leaf-feeding pests at the vegetative stage. Tissue-specific expression with the PEPC promoter allows the Cry protein to be excluded from seeds. However, Cry protein titer and insect protection dropped substantially at the flowering stage (Alinia et al., 2001) in the transgenic Bt rice with the use of the PEPC promoter. Our results with Bt rice carrying *cryIAb* driven by CaMV35S, Ubiquitin, Actin 1 and pith-specific promoters showed the rice to retain protection against stem borer and leaffolder at the flowering stage (Aguda et al., 2001). CaMV35S, Ubiquitin and Actin 1 are constitutive promoters that drive transgene expression in all tissues at all developmental stages, whereas the pith-specific promoter directs the expression in only the pith cells inside the stem where the insect larvae feeds (Datta et al., 1998).

There are still many questions that need to be addressed including a minimum requirement of Bt Cry protein to protect a given rice variety from different insect pests. We have observed that a range of 0.01-0.2 µg/g Cry protein is capable of conferring 100% yellow stem borer larvae mortality with the cut-stem bioassay (Datta et al., 1998; Datta et al., unpublished data).

CONCLUSION

Deployment of transgenic insect protected rice varieties has potential as an environment-friendly and cost-effective component in IPM. Bt rice will have a tremendous impact in Asia in controlling insect pests and reducing the financial, health, and environmental burden associated with conventional chemical pesticides. With these goals in mind, the IRRI has done extensive work to generate and test transgenic insect protected rice that will be appropriate for the growing conditions, dietary preferences and management capabilities of poor farmers in Asia.

REFERENCES

Aguda, R.M., Datta, K., Tu, J., Datta, S.K. and Cohen, M.B. (2001) Expression of *Bt* genes under control of different promoters in rice at vegetative and flowering stages. *Int. Rice Res. Notes* 26:26-27.

Alam, M.F., Datta, K., Abrigo, E., Vasquez, A., Senadhira, D. and Datta, S.K. (1998) Production of transgenic deepwater indica rice plants expressing a synthetic *Bacillus thuringiensis cryIA(b)* gene with enhanced resistance to yellow stem borer. *Plant Sci.* 135:25-30.

Alam, M.F., Datta, K., Abrigo, E., Oliva, N., Tu, J., Virmani, S.S. and Datta, S.K. (1999) Transgenic insect-resistant maintainer line (IR68899B) for improvement of hybrid rice. *Plant Cell Rep.* 18:572-575.

Alinia, F., Ghareyazie, B., Rubia, L., Bennett, J. and Cohen, M.B. (2001) Expression of effect of plantage, larval age, and fertilizer treatment on resistant of a *cryIAb*-transformed aromatic rice to lepidopterans stem borers and foliage feeders. *J. Econ. Entomol.* 93:484-493.

Baisakh, N. (2000) Improvement of rainfed lowland indica rice through *in vitro* culture and genetic engineering. Ph.D. Thesis (Utkal Univ.), India.

Balachandran, S., Chandel, G., Alam, M.F., Tu, J., Virmani, S.S., Datta, K. and Datta, S.K. (2003) Improving hybrid rices through anther culture and transgenic approaches. In: Proceedings of the 4th International Symposium on Hybrid Rice, 14-17 May 2002, Hanoi, Vietnam (in press).

Breitler, J.C., Marfà, V., Royer, M., Meynard, D., Vassal, J.M., Vercambre, B., Frutos, R., Messeguer, J., Gabarra, R. and Guiderdoni, E. (2000) Expression of a *Bacillus thuringiensis cryIB* synthetic gene protects Mediterranean rice against the striped stem borer. *Plant Cell Rep.* 19:1195-1202.

Cheng, X., Sardana, R., Kaplan, H. and Altosaar, I. (1998) *Agrobacterium*-transformed rice plants expressing synthetic *cryIA(b)* and *cryIA(c)* genes are highly toxic to striped stem borer and yellow stem borer. *Proc. Natl. Acad. Sci., USA* 95: 2767-2772.

Cohen, M.B., Savary, S., Huang, N., Azzam, O. and Datta, S.K. (1998) Importance of rice pests and challenges to their management. In: Sustainability of Rice in the Global Food System. Dowling NG, Greenfield SM, Fischer KS (eds.), Pacific Basin Study Center and IRRI, Manila, pp. 145-164.

Datta, K., Torrizo, L., Oliva, N., Alam, M.F., Wu, C., Abrigo, E., Vasquez, A., Tu, J., Quimio, C., Alejar, M., Nicola, Z., Khush, G.S. and Datta, S.K. (1996) Production of transgenic rice by protoplast, biolistic and *Agrobacterium*, systems. Proceedings of the Fifth International Symposium on Rice Molecular Biology, pp. 159-167. Yi Hsien Pub. Co. Taipei, Taiwan.

Datta, S.K., Torrizo, L., Tu, J., Oliva, N. and Datta, K. (1997) Production and molecular evaluation of transgenic rice plants. IRRI Discussion Paper Series No. 21. International Rice Research Institute, Manila 1099, Philippines.

Datta, K., Vasquez, A., Tu, J., Torrizo, L., Alam, M.F., Oliva, N., Abrigo, E., Khush, G.S. and Datta, S.K. (1998) Constitutive and tissue-specific differential expression of *CryIA(b)* gene in transgenic rice plants conferring enhanced resistance to insect pests. *Theor. Appl. Genet.* 97:20-30.

Datta, K., Vasquez, A., Khush, G.S. and Datta, S.K. (1999) Development of transgenic New Plant Type rice for stem borer resistance. *Rice Genet. Newslet.* 16:143-144.

Datta, S.K. (2000a) Potential benefit of genetic engineering in plant breeding: Rice, a case study. *Agric. Chem. Biotechnol.* 43:197-206.

Datta, S.K. (2000b) A promising debut for *Bt* hybrid rice. In: Information Systems for Biotechnology. News Report, pp. 1-3, Virginia, USA. <*http://www/biotech-info. net/promising_debut.html*> or <*www.gophisb.biochem.vt.edu/news/2000/news00.dec. html*>.

Datta, K. and Datta, S.K. (2002) Plant Transformation In: Molecular Plant Biology, Gilmartin PM and Bowler C (eds.). Vol. 1. A Practical Approach, Oxford University Press, United Kingdom, pp.13-32.

Datta, K., Baisakh, N., Thet, K.M., Tu, J. and Datta, S.K. (2002) Pyramiding transgenes for multiple resistance in rice against bacterial blight, stem borer and sheath blight. *Theor. Appl. Genet.* DOI 10.1007/s00122-002-1014-1 (online).

Frutos, R., Rang, C. and Royer, M. (1999) Managing insect resistance to plants producing *Bacillus thuringiensis* toxins. *Crit. Rev. Biotechnol.* 19:227-276.

Ghareyazie, B., Alinia, F., Menguito, C., Rubia, L.G., de Palma, J.M., Liwanag, E.A., Cohen, M.B., Khush, G.S. and Bennett, J. (1997) Enhanced resistance to two stem borers in an aromatic rice containing a synthetic *cry1Ab* gene. *Mol Breed.* 3:401-404.

Ho, N.H., Baisah, N, Oliva, N., Datta, K., Frutos, R., and Datta, S.K. (2003). Translationally focused hybrid Bt genes confer resistance to yellow stem borer (*Scirpophaga incertulas* Walker) in transgenic indica rice cultivars. *Plant Biotechnology Journal* (submitted).

Maqbool, S.B., Hunain, T., Riazuddin, S., Masson, L. and Christou, P. (1998) Effective control of yellow stem borer and rice leaf folder in transgenic rice indica varieties Basmati 370 and M7 using the novel delta endotoxin *cry2A Bt* gene. *Mol. Breed.* 4:501-507.

Marfà, V., Melé, E., Gabarra, R., Vassal, J.M., Guiderdoni, E. and Messeguer, J. (2002) Influence of the developmental stage of transgenic rice plants (cv. Senia) expressing the cry1B gene on the level of protection against the striped stem borer (*Chilo suppressalis*). *Plant Cell Report* 20:1167-1172.

Nayak, P., Basu, D., Das, S., Asitava, B., Ghosh, D., Ramakrishnan, N., Ghosh, M. and Sen, S. (1997) Transgenic elite indica rice plants expressing cry1Ac α-endotoxin of *Bt* are resistant against yellow stem borer. *Proc. Natl. Acad. Sci., USA* 94:2211-2116.

Ramaswamy, C. and Jatileksono, T. (1996) Intercountry comparison of insect and disease losses. In: Rice Research in Asia: Progress and Priorities (Evenson, R.E., Herdt, R.W., Hossain, M. eds.) pp. 305-316. CAB International in association with International Rice Research Institute, Wallingford.

Tu, J., Datta, K., Alam, M.F., Fan, Y., Khush, G.S. and Datta, S.K. (1998) Expression and function of a hybrid *Bt* toxin gene in transgenic rice conferring resistance to insect pest. *Plant Biotechnol.* 15(4):195-203.

Tu, J., Zhang, G., Datta, K., Xu, C., He, Y., Zhang, Q., Khush, G. S. and Datta, S. K. (2000) Field performance of transgenic elite commercial hybrid rice expressing *Bacillus thuringiensis* δ-endotoxin. *Nature Biotechnol.* 18:1101-1104.

Tu, J., Datta, K., Oliva, N., Zhang, G., Xu, C., Khush, G.S., Zhang, Q. and Datta, S.K. (2003) Site-independently integrated transgenes in the elite restorer rice line Minghui 63 allow removal of a selectable marker from the gene of interest by self-segregation. *Plant Biotechnology Journal* 1:155-165.

Wu, C., Fan, Y., Zhang, C., Oliva, N. and Datta, S.K. (1997) Transgenic fertile japonica rice plants expressing modified *cry1A(b)* gene resistant to yellow stem borer. *Plant Cell Reports* 17:129-132.

Ye, G. Y., Tu, J., Hu, C., Datta, K. and Datta, S. K. (2001) Transgenic IR72 with fused *Bt* gene *cry1A(b)/cry1A(c)* from *Bacillus thuringiensis* is resistant against four lepidopteran species under field conditions. *Plant Biotechnol.* 18(2):125-133.

Fulfilling the Promise of Bt Potato in Developing Countries

Marc Ghislain
Aziz Lagnaoui
Thomas Walker

SUMMARY. Potato production is expanding rapidly in developing countries, particularly in Asia. For many poor farmers potatoes represent a staple food and an important cash crop. Potato insect pests have a substantial negative impact on the livelihood of farmers. Bt potatoes have been effective on a commercial scale in controlling Colorado potato beetle in North America. Small-scale tests have demonstrated Bt's efficacy in controlling potato tuber moth. Reliance on chemical pesticides can be reduced and/or replaced by pest-protected potato, genetically engineered to express genes from the bacterium *Bacillus thuringiensis* (Bt). Reductions in insecticide use would provide significant economic, health and environmental benefits. Unintentional Bt gene escape into wild relatives of cultivated potato should be rare and unlikely to persist. However, a monitoring system should be in place in areas where sexually compatible relatives of potato occur. The potential of Bt varieties to displace land races or reduce crop genetic diversity should be addressed by the creation of sanctuaries or germplasm banks. Because farmers in developing countries exchange their seeds and mandatory labeling would likely ex-

Marc Ghislain, Aziz Lagnaoui and Thomas Walker are affiliated with the International Potato Center (CIP), Apartado 1558, Lima 12, Peru.

Address correspondence to: Marc Ghislain, International Potato Center (CIP), Apartado 1558, Lima 12, Peru (E-mail: m.ghislain@cgiar.org).

[Haworth co-indexing entry note]: "Fulfilling the Promise of Bt Potato in Developing Countries." Ghislain, Marc, Aziz Lagnaoui, and Thomas Walker. Co-published simultaneously in *Journal of New Seeds* (Food Products Press, an imprint of The Haworth Press, Inc.) Vol. 5, No. 2/3, 2003, pp. 93-113; and: *Bacillus thuringiensis: A Cornerstone of Modern Agriculture* (ed: Matthew Metz) Food Products Press, an imprint of The Haworth Press, Inc., 2003, pp. 93-113. Single or multiple copies of this article are available for a fee from The Haworth Document Delivery Service [1-800-HAWORTH, 9:00 a.m. - 5:00 p.m. (EST). E-mail address: docdelivery@haworthpress.com].

http://www.haworthpress.com/store/product.asp?sku=J153
© 2003 by Taylor & Francis.
10.1300/J153v05n02_02

clude small-scale farmers, Bt genes should be introduced only into varieties handled through the formal seed systems, providing the basis for a variety-based system for segregating genetically engineered (GE) and non-GE produce. We conclude that properly managed Bt potato technology offers substantial benefits for developing countries. *[Article copies available for a fee from The Haworth Document Delivery Service: 1-800-HAWORTH. E-mail address: <docdelivery@haworthpress.com> Website: <http://www.HaworthPress.com> © 2003 by The Haworth Press, Inc. All rights reserved.]*

KEYWORDS. Potato, genetic engineering, developing countries, agriculture, *Bacillus thuringiensis*

INTRODUCTION

For many farmers, potato is not only their food staple but also their principal cash crop. Amongst major annual crops, potato ranks first in yield, fifth in production and eleventh in cultivated area in developing countries (FAO, 2001). Over the past decade, potato production in developing countries increased steadily, by 5% per annum, while the crop's cultivated area rose by 3%. In contrast, in developed countries, yield increases were offset by a decline in area. Thus, potato is increasingly a developing country crop, with production in these countries now accounting for 45% of global production. The growth in developing country production has been especially notable in Asia: China is now the biggest producer globally, while India ranks third.

Potato productivity varies markedly among developing countries, from 2 t/ha in Swaziland to 41 t/ha in Kuwait (FAO, 2001). Both yield-limiting (water, temperature, and nutrients) and yield-reducing (pest and disease) factors explain this wide variation. Insect pests rank third after late blight and viruses in importance as yield reducers, with about 20% of the affected areas found in developing countries (Walker and Collion, 1997). Insect pests are not economically important in the main production season in China and India, but they can cause heavy economic damage in most of the other 90 developing countries where the crop is grown.

Pest losses in the field and in storage can easily reach up to 50% of total yield (Oerke et al., 1995). The use of chemical pesticides is increasing rapidly, particularly where farmers are intensifying production in order to sell to urban markets and where the crop is expanding into agro-ecological regions and planting seasons outside its traditional range. The negative impact of pesticides on human health is well documented. Indiscriminate use of pesticides, often with a total disregard for the instructions given on the label, not only leads to poor pest control but also exposes both farmers and consumers to unnecessary health risks, and damages the environment. Pesticide use has been held responsible for causing a range of adverse effects on farm workers, diminishing their efficiency as producers. In Carchi, Ecuador, an area with intensive potato production all year, farmers routinely handle all classes of highly toxic pesticides with minimal care and no protective equipment, resulting in chronic dermatitis, reduced vibration sensations and significant neuropsychological disorders (Crissman et al., 1994; Cole et al., 1998). Because of illness, production costs rose, work capacity decreased, and decision-making ability eroded over time of exposure, throughout the life of the individual.

However, pesticides are still absolutely indispensable in potato production. Biological pesticides, thought to be a safe alternative, have yet to be developed and deployed on a commercial scale for the major pests of potato. A recent analysis of organic potato production in Europe revealed that the use of pesticides is unavoidable, given the lack of genetic resistance to pests and diseases in commonly grown varieties (Tamm, 1999). A similar analysis was conducted in the US, where pesticide use was again found to be unavoidable if the availability, quality, and low price of potatoes were to be maintained (Guenthner et al., 1999). The tradeoff between improving human health and reducing potato production as a consequence of reducing pesticide use can be avoided in the long term by replacing the varieties currently grown with new ones that have enhanced genetic protection from to pests and diseases.

INSECT PESTS OF POTATO

Different Pests in Different Agro-Ecological Zones

In developing countries of the tropics and subtropics, several insect pests infest potato crops with varying levels of severity. At the International Potato Center (CIP), the priority pests for research are potato tuber moths, Andean weevils, leafminer flies and aphids. The major insect pest of temperate regions, Colorado potato beetle, *Leptinotarsa decemlineata* (Say) (Coleoptera: Chrysomelidae), is not known to be a pest in the tropics.

The potato tuber moth (PTM) is the most damaging lepidopteran pest of potato in the tropics and subtropics (Lagnaoui et al., 2001). The PTM complex consists of three species: *Phthorimaea operculella* (Zeller), *Symmetrischema tangolias* (Gyen), and *Tecia solanivora* (Polvony) (all Lepidoptera: Gelechiidae). *P. operculella* is the most widely distributed of the three, attacking potato foliage, stems, and tubers, although field damage seldom causes yield losses. PTM is the single most significant insect pest of potato (field and storage) in North Africa and the Middle East (Fuglie et al., 1992). In the absence of cold stores, farmers must rely on chemical insecticides to protect tubers from tuber moth damage during the storage period (Roux et al., 1992). Storage losses reported from India vary from 30 to 70% (Saxena and Rizvi, 1974). *S. tangolias* is considered a serious pest throughout the Andean region, causing significant economic losses in potato stores in Peru and Bolivia. *T. solanivora,* an increasingly important species causing severe damage in storage, has spread from Guatemala to Venezuela and Colombia and has recently been reported in southern Ecuador. Bracing for the onslaught, Peru has launched a large-scale detection program, which is working to develop a regional integrated pest management (IPM) strategy that will give farmers better access to the tools and knowledge needed to fight the pest. One of the tools considered is the use of transgenic potatoes.

The Andean potato weevils are the most serious pests indigenous to the high altitudes of the Andes, spreading from northern Argentina, through Bolivia, Peru, Ecuador and Colombia, to Venezuela. Several species comprise what is often referred to as the Andean potato weevil complex. Most are in the genus *Premnotrypes* (Coleoptera: Curculionidae), the major species of which are *Premnotrypes latithorax* (Pierce), *Premnotrypes suturicallus* (Kuschel), and *Premnotrypes vorax* (Hustache). Other species of importance are *Rhigopsidius tucumanus* (Heller) (Coleoptera: Curculionidae), which is limited to Argentina and the south of Bolivia, and *Phyrdenus muriceus* (Germar) (Coleoptera: Curculionidae), known to attack potato and other solanaceous crops in these regions. Infestation by potato weevils can affect up to 85% of tubers. CIP has promoted a number of IPM measures suitable for use by small-scale farmers (Cisneros and Gregory, 1994). Despite the heavy use of insecticide in the Andes, the weevils are still causing severe losses. A low-cost, effective control measure will be particularly attractive to poor farmers, relieving them of the high costs and undesirable effects of toxic insecticides.

The Colorado potato beetle has the greatest potential to reduce yields in the temperate regions. The destructive, defoliating nature of this pest alarms potato growers and often results in unnecessary insecticidal sprays. Overuse of all major classes of chemicals has resulted in wide-spread insecticide resistance (Radcliffe and Lagnaoui, 1990). Although a pest of the temperate areas of the north, the Colorado potato beetle has recently been moving south and is now a serious pest in Turkey, Romania, and Albania. It has also established a significant presence in the potato growing area of northern China. The insect may well continue to adapt to more southern areas, especially in the light of the general increase in pest population movements caused by factors such as international trade and global climate change (Jeffree and Jeffree, 1996).

Genetic Engineering and Integrated Pest Management

The economic loss caused by potato pests is great due to both high damage levels and high cost of (chemical) control measures. Chemical control, heavily relied upon by farmers, is seldom effective under small potato farm conditions, due to erroneous application with regard to both types of insecticides applied and application methods. The relatively cheaper broad-spectrum insecticides are used, often in cocktails of several insecticides and fungicides, causing eradication of natural enemies of pests and development of insecticide resistance in pest populations.

The International Potato Center, in collaboration with several institutions of Latin America, has developed technologies and implemented projects on integrated management of key potato pests. These technologies were sought not only to reduce the production costs associated with pesticides but also to improve the health of agricultural workers and consumers. This was achieved by maintaining potato pest populations at economically acceptable levels by increasing the predictability and thereby the effectiveness of IPM components (Cisneros and Gregory, 1994). In the case of PTM in North Africa (Tunisia and Egypt), a sound IPM strategy, based on the use of pheromone traps, cultural practices and safer biopesticides, has all but eliminated the need for toxic chemical pesticides (Fuglie, 1995; Lagnaoui et al., 1996).

The IPM strategy for both pest complexes hinges on the availability and efficacy of locally produced biological insecticides. However, their field applications are not cost-effective in controlling the pests. Under storage conditions, Granulosis virus-based biopesticide gives satisfactory PTM control while the efficacy of the Beauveria-based biopesticide for weevil control is variable and inconsistent (Winters et al., 1997). In both cases, availability of quality products is still the weakest link as most suppliers are public-sector research institutes or development agencies. The private sector has so far proved reluctant to invest, primarily because demand fluctuates substantially from year to year in accordance with pest infestation and also because the product is bulky and difficult to transport.

The development of protected/tolerant cultivars is a very important component of an efficient IPM strategy for smallholders. Unfortunately in the case of potato tuber moth, no reliable source of resistance has been found in potato germplasm collections (Raman, 1994). The use of biotechnological approaches to introduce Bt insecticidal genes into commercial potato varieties thus holds promise as a key component in protecting potato against PTM. Moreover, conventional varietal development is slow in potatoes, a factor that further increases the potential value of a transgenic solution. A Bt potato should figure prominently as one of the components in an IPM program.

ENGINEERING PROTECTION AGAINST INSECT PESTS

Genetic Engineering in Potato

Potato is particularly difficult to breed using conventional techniques due to its heterozygous nature, tetrasomic inheritance, and inbreeding depression. Varietal development can take decades, depending on the source of germplasm and the requirements of the market. In contrast, *Agrobacterium*-mediated transformation of potato is rapid and relatively easy, because the crop lends itself to tissue culture and clonal propagation.

Potato was the first of the world's staple crops to be genetically engineered, almost 15 years ago, with the first generation of commercial genetically engineered (GE) cultivars possessing improved virus resistance (Ooms et al., 1987; Stokhaus et al., 1987; Ghislain et al., 1998, 1999). In developing countries, potato varieties were engineered for resistance to viruses and bacteria, or for improved nutritional value. This work was done in university laboratories in collaboration with national agricultural research programs in Venezuela, Argentina, Chile, Uruguay, Brazil, Colombia, and Peru, with support from CIP. More recently, other developing countries have successfully transformed potato, including Egypt, South Africa, India, and China (CIP, unpublished; Douches, pers. comm.).

Field trials on GE crop varieties have been conducted since 1986. Potato was again among the pioneer species, with a first field trial being conducted in New Zealand in 1988 (James and Krattinger, 1996). Developing countries were much slower to test GE potatoes, with a first field trial taking place in Mexico in 1992 (James and Krattinger, 1996; Frederick et al., 1995). The interest in conducting field trials in developing countries is increasing. However, so too is the concern that the human capacity for running these trials appropriately has not yet been developed, despite numerous international initiatives such as the Cartagena Protocol on Biosafety, a project on the development of national biosafety frameworks supported by the Global Environment Fund (GEF) of the United Nations Environment Programme (UNEP), and the efforts of international agencies such as the International Center for Genetic Engineering and Biotechnology (ICGEB), the Food and Agriculture Organization of the United Nations (FAO), and the centers of the Consultative Group on International Agricultural Research (CGIAR).

GE potato varieties with insect protection, herbicide tolerance, virus resistance, and modified starch have been commercially available for some time, and the first variety containing the Bt gene was released in 1996. However, the area cultivated to these varieties remains small, barely exceeding 4% overall. This apparent stagnation may change as additional developing countries open up to GE technology.

THE Bt TECHNOLOGY IN POTATO

Bt is used as a bio-insecticide in potato on account of its insecticidal crystal proteins (ICPs). The transfer of genes encoding ICPs into potato proved effective in protecting the crop against both lepidopteran (Perferoen et al., 1990) and coleopteran (Adang et al., 1993; Perlak et al., 1993) pests. Bt genes and proteins follow the nomenclature of Crickmore et al. (1998).

Potatoes with protection against PTM were first obtained using the *cry1Ab5* gene, previously *cry1Ab*, in European varieties (Perferoen et al.,1990). Improved expression was obtained through modification of the gene structure and the rate of protection was increased from an average of 10% to approximately 90% (Van Rie et al., 1994; Jansens et al., 1995). This Bt gene was transferred to CIP in Peru, where it was used to transform some 10 potato varieties adapted to different agro-ecological zones in developing countries where PTM is an important constraint (Cañedo et al., 1999).

The Cry2Aa1, Cry1Ba1 and, to a lesser extent, Cry1C ICPs have also been reported to be toxic to PTM, but no transgenic potatoes expressing them have yet been developed (Van Rie et al., 1994; Bradley et al., 1995). A *cry1Ac* gene has been introduced to potato and shown to cause up to 10% mortality of first-instar PTM larvae (Ebora et al., 1994). The *cry1Ia1* gene (previously *cryV*) was more recently engineered for increased expression in plants and transferred to potato varieties well adapted to temperate and sub-temperate climates (Douches et al., 1998; Mohammed et al., 2000). Protection from the two tuber moth pests, *P. operculella* and to *S. tangolia*, was obtained using either *cry1Ab5* and *cry1Ia1* genes, with the former being slightly more effective (Lagnaoui et al., 2001). Lastly, another Bt gene, *cry9Aa1*, was introduced into a Finnish potato variety and shown to confer protection from PTM at CIP (Kuvshinov et al., 2001; Lagnaoui, unpublished).

The availability of at least seven different Bt genes (*cry1Ab5*, *cry1Ac*, *cry1Ba1*, *cry1Ia1*, *cry1C*, *cry2Aa1*, and *cry9Aa1*) to protect potato offers the potential of developing powerful resistance management strategies that have unfortunately not been exploited so far. These include gene pyramiding (see Sharma et al., this volume, pp. 53-76) and rotation of insecticide modes of action.

Field trials of Bt potatoes with protection against PTM have so far been conducted in very few developing countries: Peru, Tunisia, Egypt, and recently in South Africa. In Peru, nine field trials were conducted between 1994 and 1998 at five different locations with three different varieties expressing the *cry1Ab5* gene (CIP, unpublished). These trials allowed the evaluation of field protection at foliar level against PTM and the identification of differences in plant structure, as reported by Cañedo et al. (1999). Field trials in Tunisia were conducted in 1995 using potato expressing the *cry1Ab5* gene; however, this work was not completed and results were reported only at a conference (Khamassy and Ben Salah, 1996). In Egypt, two potato varieties, Spunta and a breeding clone, expressing the *cry1Ia1* gene were field-tested (Douches, pers. comm.). This material was also more recently field-tested in South Africa and evaluated under greenhouse conditions in Indonesia where at least one line was similar in productivity to the non-Bt counterpart of the commercial variety Spunta (Douches, pers. comm.). Therefore, Bt potatoes transformed for protection against PTM have passed the first steps of 'proof-of-utility.' Today, medium- to large-scale field trials are needed to select the best performing Bt potato lines and to validate management strategies that will prevent the development of insect resistance and Bt gene escapes.

In developing countries, the most important coleopteran pest is the Andean potato weevil. A recombinant Cry3Aa protein has been found effective against this pest (Gomez et al., 2000). However, no transgenic potatoes have yet been successfully engineered against this pest, despite its importance and the lack of cost-effective and environmentally friendly control measures.

In the northern temperate zone, a modified *cry3Aa1* gene has been used to enhance protection of the Russet Burbank variety against the destructive Colorado potato beetle (Adang et al., 1993; Perlak et al., 1993). Another *cry3* gene, *cry3Ca1*, was found to be effective against this pest and was engineered for enhanced insecticidal activity (Haffani et al., 2000). Other *cry* genes for Cry1 ICPs have also been found effective (Naimov et al., 2001). The value of potatoes engineered for resistance to Colorado potato beetle is limited for developing countries to those in temperate zones. Such potatoes could be of great value to countries of the former eastern block because of their weak seed systems and somewhat limited access to chemical inputs.

ENVIRONMENTAL SAFETY

Effects on Non-Target Organisms

Commercial fields of New Leaf™ potatoes with Bt-based protection against Colorado potato beetle were evaluated for impact on non-target organisms leading to two main conclusions: (1) beneficial arthropods were unaffected and could help to control secondary pests; and (2) beneficial insects and spiders were more abundant in the New Leaf fields (Betz et al., 2000).

The risk to non-target organisms of Bt-pollen from GE potato appears to be negligible. This assessment is based on the observation that: wind pollination is only occasionally observed at short distances (White, 1983); that flower-visiting insects are almost exclusively bumble bees, which belong to an insect family not affected by this Bt ICP; and that potato flowers do not produce nectar and hence are not attractive to other insects.

The effect of Cry3A ICP, at a sublethal dose, on the susceptibility of Colorado potato beetle to the entomopathogen *Beauveria bassiana* (Balsamo) Vuillemin (Deuteromycota: Hyphomycetes) was evaluated (Costa et al., 2001). The authors demonstrated that, even though the development of resistance to Bt ICPs might negatively impact other biological control methods, such an event was unlikely to take place.

Nevertheless, testing the effect of Bt genes on non-target species and in particular beneficial insects is an important step in the safety assessment of these GE potatoes. The *cry1Ab5* potatoes have been tested in this respect and found to be protected against the tomato pinworm, *Tuta absoluta* Meyrick (Lepidoptera: Gelechiidae), while other pests were unaffected (Cañedo et al., 1999). Due to the specificity of Bt ICPs, the technology is atypically safe for non-target organisms, among available pesticides (see Federici, this volume, pp. 11-30).

Durability of Pest Protection

The process of deploying Bt potato in developing countries needs to be carefully examined with regard to pest management. Resistance to Bt has been examined in several crop pests, such as diamondback moth (*Plutella xylostella* [L.] [Lepidoptera: Plutellidae]), Indian meal moth (*Plodia interpunctella* Hübner [Lepidoptera: Pyralidae]), tobacco budworm (*Heliothis virescens* F. [Lepidoptera: Noctuidae]) and Colorado potato beetle, and was recently reviewed (McGaughey and Whalon, 1992; Tabashnik, 1994; Ferré and Van Rie, 2002). Gould (1996) categorized the various tactics to counter development of resistance into five strategies aiming at decreasing adaptedness, fitness or both. The high-dose refuge model of resistance management is presented in detail by Matten and Reynolds (this volume, pp. 137-178) and a number of other strategies are discussed by Sharma et al. (this volume, pp. 53-76).

Developing country agro-ecologies are extremely diverse and no single strategy can be designed *a priori* without considering the context in which it will be applied. In general, Bt potatoes with protection against PTM will be of interest to farmers in two situations: (1) where PTM is an important pest in both field and storage; and (2) where storage is needed to commercialize the produce.

Gene Flow

In this review, gene flow is understood in a general sense of the escape of transgenes from their original context. The flow of transgenes to non-plant organisms is not considered here because it has been shown experimentally to be irrelevant as a new risk posed by transgenic crops (Schlüter et al., 1995). Gene flow relates to volunteers, to the possibility of naturalization, to the production of hybrids and their fate, and to the establishment of a weedy population. Gene flow is perceived as posing risks of two kinds: transgene introgression into cultivated varieties and genetic 'contamination' of land races or wild relatives. These risks need separate treatment as they raise very different issues.

Gene Flow in Cultivated Varieties

Cultivated varieties of potato are essentially derived from the tetraploid species *Solanum tuberosum* L. (Solanaceae) which has two subspecies: *tuberosum*, well-adapted to long-day conditions and *andigena*, well-adapted to short-day conditions. Many modern varieties of *tuberosum* origin are sterile through a number of distinct genetic systems, reducing significantly seed sets. In addition, potato flowers, devoid of nectar, are visited by a limited range of insects, such as bumble bees, that are not ubiquitously distributed (Hanneman, 1995). Transgene flow from GE potato varieties has also been tested and found to be insignificant beyond a distance of 20 m (McPartlan and Dale, 1994). Elsewhere, transgene flow has been reported at a distance of more than 1000 m (Skogsmyr, 1994; Conner and Dale, 1996), though the pollen vector was an unusual pollinator insect. Preliminary experiments in Bolivia demonstrate gene flow beyond the 20 m distance (Bravo et al., 2002).

Cross-pollination between transgenic and non-transgenic modern varieties, though possible in principle, will not likely result in transgene introgression for two reasons. Firstly, farmers do not collect potato berries and their remains are eliminated by farmers' practices before the potato tubers are harvested. Farmers only collect berries in true potato seed production but for this purpose isolation systems have already been established to prevent gene flow from neighboring fields. Secondly, farmers maintain variety uniformity because of the pressure from the market and also because they use certified seed. Seed tubers are purchased through seed systems that ensure quality and purity of the genetic material. Establishment of volunteers derived from transgenic potato cultivation was shown to be negligible, indeed less than from a traditionally bred variety, in a study conducted over a 10-year period in the UK (Crawley et al., 2001). Gene flow between transgenic and non-transgenic cultivated potato varieties is an event that will rarely occur naturally and will not persist under normal production practices.

Gene Flow in Land Races or Wild Relatives

From the southern US to Chile and Argentina, a continuum of native potatoes and wild species are present, comprising a total of seven cultivated potato species of four different ploidy levels and inter-specific hybrids with undefined taxonomic unit affiliation, as well as 199 wild species; in total these provide numerous opportunities for sexual crossing (Spooner and Hijmans, 2001). Outside this American band, no sexually-compatible relatives are present in potato growing areas (Hijmans and Spooner, 2001). Therefore, the issue of gene flow through pollination into potato relatives is relevant only in its centers of origin and diversity.

Land races in the Andean region are characterized by a wide diversity of varieties and cultivars, which have been kept by farming communities and are constantly intercrossed by open pollination (Quiros et al., 1992). Gene flow to wild species has been documented even though most of these species are rare and endemic (Spooner et al., 1999). Hybridization between wild and cultivated potato has apparently occurred in the past and has been documented for a number of cases (Spooner and Hijmans, 2001). For example, the most likely origin of the Bolivian weedy potato species *Solanum sucrense* Hawkes is from hybridization between *S. tuberosum* subsp. *andigena* and *Solanum oplocense* Hawkes (Astley and Hawkes, 1989). Similarly, a field experiment conducted in the Peruvian highlands demonstrated that botanical seeds are often hybrid seeds with wild species parentage (Rabinowitz et al., 1990).

The debate over gene flow in centers of origin and diversity is polarized between the 'contamination' position and the 'improvement' position. The former presupposes that the presence of a transgene outside its original variety or target agricultural environment is a negative event and hence should be considered pollution. The latter presupposes that the escape of the transgene will carry over its benefits into the traditional variety or agricultural environment and hence represents an improvement for farmers. Ultimately it will require a rigorous scientific assessment involving field trials and surveys of grower practices, to determine if there will be any positive or negative effects on crop genetic diversity.

The biological issues relating to transgene flow have already been discussed (Jackson and Hanneman, 1996; Frederick et al., 1995). It has been proposed that the region in which crossing can occur can be divided into areas, according to level of risk, using species inventory and geographical information system techniques, as a means to concentrate on site-specific cases (Hijmans, pers. comm.). These studies indicate that gene flow will occur in the long run in very specific areas, as is the case for land races. Gene flow *per se* is not a risk: the main concern regarding the transfer of Bt genes lies in its potential negative impact on potato genetic diversity.

The particular case of resistance to insect pests mediated by Bt genes does not appear to represent a potential disruptive force to gene pools of wild relatives. The insects targeted by Bt transgenes are thought not to be the limiting constraints on wild populations. Hybrids between cultivated varieties and wild relatives are likely to be less adapted to the harsh conditions that prevail in the natural habitat of the wild parent, and a Bt gene would not provide a selective advantage to compensate for such fitness loss. This makes it seem unlikely that a Bt trait could present a selective advantage that would swamp out genetic diversity, or create a heartier weed in populations of wild relatives, but this remains to be demonstrated through experimentation.

The occurrence of hybrids with land races carrying Bt genes may, in contrast, be of concern in specific cases where a particular pest is difficult to manage and is present in the area of cultivation of native potatoes. This is not the case for the Colorado potato beetle, nor probably for PTM, but certainly is of concern for the Andean potato weevil. It is likely that farmers would be inclined to breed a Bt transgene conferring resistance to the weevils into their land races. Although pest management in the land races would benefit, such a development might be conducive to genetic erosion, because some land races might not hybridize with the Bt variety and would thus come to be at a disadvantage. To date, no such Bt gene technology is available but the demand for it exists and potential Bt ICPs have been recently identified (Gomez et al., 2000). Hence, in circumstances under which farmers would be ready to abandon the conservation of some of their land races in favor of a weevil-resistant Bt potato, steps have to be taken to ensure that the genetic diversity of land races is conserved *in situ* or elsewhere.

THE ECONOMICS OF Bt POTATOES

The Fate of Bt Potatoes in the US

Beginning in 1995, certified seed of Bt potato was available to growers in the US and Canada from Naturemark®, the potato subsidiary of Monsanto. The Bt potato was supplied in several varietal backgrounds and was targeted at contributing solutions to the problem of the Colorado potato beetle, the most economically damaging pest of the crop in North America. In spite of the economic importance of the pest, the area planted to Bt potato did not exceed 4% of that planted to certified seed in any year during the 1990s. This is likely due in part to the existence of the effective foliar insecticide control imadicloprid. The area of Bt potato peaked at about 3,000 acres in 1996. In 2000, one potato processing company, in response to pressure from a fast food chain, decided not to purchase any transgenic material for processing. This decision was soon followed by others in the industry and led to the closure of Naturemark in 2001. This event, though understandable from a marketing perspective, may well complicate the near-term future of the Bt potato for other countries and regions where pests are causing severe damage and pesticide use is problematic.

Barriers to Deployment of Bt Potatoes

The national agricultural research systems of developing countries, as well as the international agricultural research centers such as CIP, have been slow to invest in evaluating the prospects of applying transgenic plant technologies to potato. There are at least three reasons for this. First, technical progress has been slow. Second, biosafety protocols in developing countries have been mired in some cases by the absence of infrastructure for their management, and in others by a severely conservative design, resulting in substantial delay to field-testing (Paarlberg, 2000). Third, a study on the expected benefits of GE potato in developing countries had already been carried out which made additional *ex-ante* assessment studies less needed (Qaim, 1998). Because the Bt potato technology has controversial aspects in the Andean region, such as the potential impact of transgenic varieties on biodiversity and on poor farmers' livelihoods, decision- and policy-makers are awaiting supporting evidence of the likely benefits accruing from the deployment of these varieties. Hence, more research is needed to document the potential for positive or negative impact of Bt potato in well-defined scenarios of potato production.

Bt-mediated protection is now beginning to play a role in agriculture in developing countries, and the early evidence of impact is largely positive for rural livelihoods and for the environment (Huang et al., 2001; Traxler et al., 2001; Ismael et al., 2001). The economic importance of PTM is concentrated in North Africa, the Middle East, and in the Andes. These regions thus stand to gain in productivity and reduced insecticide use through deployment of Bt potatoes. Egypt is the first country where the national research system has developed field testing on research stations (Mohammed et al., 2000; Douches, pers. comm.).

However, the benefit of production gains can be counteracted by changes in the marketability of any potential exports. About 10% of the total crop production in Egypt is exported. It may be prohibitively costly to segregate the domestic and export markets as the same varieties dominate both markets, so commercial production of the Bt potato in North African and Middle Eastern countries may have to wait until it is formally approved for consumption in Europe. For this reason, the prospects may be brighter for the Bt potato in Andean countries where *T. solanivora* is becoming a major threat to potato production.

Deployment of GE potatoes, and in particular Bt potatoes, is often perceived as constrained by numerous patents. Indeed there are many patents covering Bt technology and *Agrobacterium*-mediated transformation. However, it is a misconception to interpret these as barriers to deployment. Licensing for commercial use is possible, although constrained by liability issues. Free licensing to resource-poor farmers is another possibility. At CIP, the Bt potato can be deployed under a non-exclusive royalty-free license in developing countries. However, public domain Bt genes and genic elements are also available. In addition, most of the early patents in agricultural biotechnology were limited to developed countries and therefore copying the technology and deploying it for commercial purposes in countries where these patents have not been registered is perfectly legal. So long as produce is not exported, intellectual property rights do not appear to be a serious constraint to the adoption of Bt potatoes in developing countries.

Seed potatoes are small tubers obtained either from the previous harvest or from a seed supplier. The first case, referred to as the informal seed system, is often limited to land races–farmers' varieties grown for home- or community-consumption. The formal seed system, on the other hand, supplies planting material to farmers growing potatoes for either the processing or fresh table markets. Only the formal seed system can guarantee the genetic uniformity of seed that is a pre-condition for uniform produce. Therefore, the restriction of GE technology to varieties within the formal seed system offers the only possibility of achieving a cost-effective segregation between GE and non-GE potatoes. This strategy, applicable in developing countries, could still be to the benefit of resource-poor farmers since they also grow commercial varieties for the market. It does, however, present logistical and regulatory challenges that may be prohibitive for developing countries seeking to export produce to markets where segregation is required.

CONCLUSIONS

Pest-protected potatoes can be generated using transgenic Bt technology. Their deployment offers potential economic, health and environmental benefits with increased yields and reduced pesticide use. Bt transgene escape into, followed by persistence in, wild relatives of potato is regarded as an extremely remote scenario. However, farmers may want to introgress specific Bt genes into land races for protection against Andean potato weevils, posing a potential threat to land race diversity in centers of origin and diversity. Measures should be taken to preserve existing crop genetic diversity. As long as market forces favor segregation, Bt genes should be introduced only into varieties handled through the formal seed system in developing countries, providing the basis for a variety-based system for segregating GE and non-GE produce, and protecting small, poor farmers from exclusion in export markets that demand segregation.

REFERENCES

Adang, M.J., Brody, M.S., Cardineau, G., Eagan, N., Roush, R.T., Shewmaker, M.K., Jones, A., Oakes, J.V., and BcBride, K.E. (1993) The reconstruction and expression of a *Bacillus thuringiensis cry IIIA* gene in protoplasts and potato plants. *Plant Molecular Biology* 21:1131-1145.

AGBIOS. 2001. Essential Biosafety database. *<http://www.agbios.com>*.

Astley, D. and Hawkes, J.G. (1979) The nature of the Bolivian weed potato species *Solanum sucrense* Hawkes. *Euphytica* 28(3):685-696.

Betz, F.S., Hammond, B.G., and Fuchs, R.L. (2000) Safety and advantages of *Bacillus thuringiensis*-protected plants to control insect pests. *Regulatory Toxicology and Pharmacology* 32:156-173.

Bradley, D., Harkey, M.A., Kim, M.-K., Biever, K.D., and Bauer, L.S. (1995) The insecticidal CryIB crystal of *Bacillus thuringiensis* ssp. *thuringiensis* has dual specificity to coleopteran and lepidopteran larvae. *Journal of Invertebrate Pathology* 65:162-173.

Bravo, W., Franco, J., Main, G., Carrasco, E. and Gabriel, J. (2002) Evaluación de la dispersion de polen como medida de bioseguridad para la liberación de plantas transgénicas de papa a campo. *Revista Latinoamerican de la Papa* 13:95-103.

Cañedo, V., Benavides, J., Golmirzaie, A., Cisneros, F., Ghislain, M., and Lagnaoui, A. (1999) Assessing Bt-transformed potatoes for potato tuber moth, *Phthorimaea operculella* (Zeller) management. *In*: Impact on changing a world, Program report 1997-98. International Potato Center, pp. 161-169.

Cisneros, F. and Gregory, P. (1994) Potato pest management. *Aspects of Applied Biology* 39:113-124.

Cole, D.C., Carpio, F., Julian, J.A., and Leon, N. (1998) Health impacts of pesticide use in Carchi farm populations, pp. 209-230. *In*: Crissman, C.C., Antle, J.M., and Capalbo (eds.), Economic, environmental, and health tradeoffs in agriculture: pesticides and the sustainability of Andean potato production. Kluwer Academic Publishers. Massachusetts, USA. pp. 209-230.

Conner, A.J. and Dale, P.J. (1996) Reconsideration of pollen dispersal data from field trials of transgenic potatoes. *Theoretical and Applied Genetics* 92:505-508.

Costa, S.D., Barbercheck, M.E., and Kennedy, G.G. (2001) Mortality of Colorado potato beetle (*Leptinotarsa decemlineata*) after sublethal stress with the CryIIIA δ-endotoxin of *Bacillus thuringiensis* and subsequent exposure to *Beauveria bassiana*. *Journal of Invertebrate Pathology* 77:173-179.

Crawley, M.J., Brown, S.L., Hails, R.S., Kohn, D.D., and Rees, M. (2001) Transgenic crops in natural habitats. *Nature* 409:682-683.

Crickmore, N., Zeigler, D.R., Feitelson, J., Schnepf, E., Van Rie, J., Lereclus, D., Baum, J., and Dean, D.H. (1998) Revision of nomenclature for the *Bacillus thuringiensis* pesticidal crystal proteins. *Microbiology and Molecular Biology Reviews* 62(3):807-813.

Crissman, C.C., Cole, D.C., and Carpio, F. (1994) Pesticide use and farm worker health in Ecuadorian potato production. *American Journal of Agricultural Economics* 76:593-597.

Douches, D.S., Westedt, A.L., Zarka, K., and Schroeter, B. (1998) Potato transformation to combine natural and engineered resistance for controlling tuber moth. *HortScience* 33(6):1053-1056.

Ebora, R.V., Ebora, M.M., and Sticklen, M.B. (1994) Transgenic potato expressing the *Bacillus thuringiensis* CryIA (c) gene effects on the survival and food consumption of *Phthorimaea operculella* (Lepidoptera: Gelechiidae) and *Ostrinia nubilalis* (Lepidoptera: Noctuidae). *Journal of Economic Entomology* 87(4):1122-1127.

FAO. (2001) FAOSTAT database, <*http://apps.fao.org*>.

Ferré, J. and Van Rie, J. (2002) Biochemistry and genetics of insect resistance to *Bacillus thuringiensis*. *Annual Review of Entomology* 47:501-533.

Frederick, R.J., Virgin, I., and Lindarte, E. (eds.). (1995) Environmental concerns with transgenic plants in centers of diversity: Potato as a model. Proceedings of a regional workshop, Parque Nacional Iguazú, Argentina, 2-3 Jun 1995. Biotechnology Advisory Commission (BAC) Inter-American Institute for Cooperation on Agriculture (IICA). p. 70.

Fuglie, K., Ben Salah, H., Essamet, M., Ben Temime, A., and Rahmouni, A. (1992) The development and adoption of integrated pest management of the potato tuber moth, *Phthorimaea operculella* (Zeller) in Tunisia. *Insect Science and Its Application* 14(4):501-509.

Fuglie, K. (1995) Measuring welfare benefits from improvements in storage technology with an application of Tunisian potatoes. *American Journal of Agricultural Economics* 77(1):162-173.

Ghislain, M. and Golmirzaie, A. (1998) Genetic engineering for potato improvement, pp. 115-162. *In*: Khurana, P., Chandra, R., Upadhya, M. (eds.), Comprehensive Potato Biotechnology. MPH Publishing House, New Delhi, India.

Ghislain, M., Bonierbale, M., and Nelson, R. (1999) Gene technology for potato in developing countries, pp.135-140. *In*: Hohn, T., and Leisinger, K.M. (eds.), Biotechnology of Food Crops in Developing Countries. Springer Wien, NewYork, USA.

Gomez, S., Mateus, A.C., Hernandez, J., and Zimmermann, B.H. (2000) Recombinant Cry3Aa has insecticidal activity against the Andean potato weevil, *Premnotrypes vorax*. *Biochemical and Biophysical Research Communications* 279(2):653-556.

Gould, F. (1996) Deploying pesticidal-engineered crops in developing countries, pp. 264-293. *In* Persley, G. (ed.), Biotechnology and Integrated Pest Management. CAB International, Wallingford, UK.

Guenthner, J.F., Wiese, M.V., Pavlista, A.D., Sieczka, J.B., and Wyman, J. (1999) Assessment of pesticide use in the US potato industry. *American Journal of Potato Research* 76:25-29.

Haffani, Y.Z., Overney, S., Yelle, S., Bellemare, G., and Belzile, F.J. (2000) Premature polyadenylation contributes to the poor expression of the *Bacillus thuringiensis cry3Ca1* gene in transgenic potato plants. *Molecular and General Genetics* 264(1-2): 82-88.

Hanneman, R.E. (1995) Ecology and reproductive biology of potato: the potential for and environmental implications of gene spread, pp. 19-38. *In*: Frederick R.J., Virgin, I., and Lindarte, E. (eds.), Environmental Concerns with Transgenic Plants in Centers of Diversity: Potato as a Model. Proceedings of a regional workshop, Parque Nacional Iguazú, Argentina, June 2-3, 1995.

Hijmans, R.J. and Spooner, D.M. (2001) Geographic distribution of wild potato species. *American Journal of Botany* 88(11):2101-2112.

Huang, J., Hu, R., Pray, C., Qiao, F., and Rozelle, S. (2001) Biotechnology as an alternative to chemical pesticides: a case study of Bt cotton in China. Electronic Proceedings of the 5th International Conference on Biotechnology, Science and Modern Agriculture. International Consortium on Agricultural Biotechnology Research (ICABR) held in Ravello, Italy. June 15-18, 2001 (*http://www.economia.uniroma2.it/ conferenze/icabr01/Program.htm*).

Ismael, Y., Beyers, L., Thirtle, C., and Piesse, J. (2001) Efficiency of Bt cotton adoption by smallholders in Makhathini Flats, Kwazulu-Natal, South Africa. Electronic Proceedings of the 5th International Conference on Biotechnology, Science and Modern Agriculture. International Consortium on Agricultural Biotechnology Research (ICABR) held in Ravello, Italy. June 15-18, 2001 (*http://www.economia. uniroma2.it/conferenze/icabr01/Program.htm*).

Jackson, S.A. and Hanneman, R.E. (1996) Potential gene flow between cultivated potato and its wild tuber-bearing relatives: Implications for risk assessment of transgenic potatoes. Proceedings of the Biotechnology Risk Assessment Symposium held from June 23 to 25, 1996, Ottawa, Ontario, Canada.

James, C. and Krattinger, A.F. (1996) Global review of the field-testing and commercialization of transgenic plants, 1986 to 1995: the first decade of crop biotechnology. *ISAAA Briefs* No. 1. International Service for the Acquisition of Agri-biotech Applications (ISAAA), Ithaca, NY, USA. 31 pp.

Jansens S., Cornelissen, M., De Clercq, R., Reynarts, A., and Peferoen, M. (1995) *Phthorimaea operculella* (Lepidoptera: Gelechiidae) resistance in potato by expression of the *Bacillus thuringiensis* CryIA(b) insecticidal crystal protein. *Journal of Economic Entomology.* 88(5):1469-1476.

Jeffree, C.E. and Jeffree, E.P. (1996) Redistribution of the potential geographical ranges of mistletoe and Colorado beetle in Europe in response to the temperature component of climate change. *Functional Ecology* 10(5):562-577.

Khamassy, N. and Ben Salah, H. (1996) Evaluation agronomique et entomologique de clones transgéniques de pomme de terre resistants à la teigne *Phthorimaea operculella* Zeller. Proceedings of 13 Triennial Conference of the European Association for Potato Research. Veldhoven (Netherlands). 14-19 Jul 1996. pp. 625-626.

Kuvshinov, V., Koivu, K., Kanerva, A., and Pehu, E. (2001) Transgenic crop plants expressing synthetic *cry9Aa* gene are protected against insect damage. *Plant Science* 160:341-353.

Lagnaoui, A., Ben Salah, H., and El Bedewy, R. (1996) Integrated Management to Control Potato Tuber Moth in North Africa and the Middle East. *CIP Circular* 22(1):10-15.

Lagnaoui, A., Cañedo, V., and Douches, D.S. (2001) Evaluation of *Bt-cry1Ia1* (*cryV*) transgenic potatoes on two species of potato tuber moth, *Phthorimaea operculella* (Zeller) and *Symmetrischema tangolias* (Gyen) in Peru, pp. 117-121. *In*: Program Report 1998-2000. International Potato Center, Lima, Peru.

McGaughey, W.H. and Whalon, M.E. (1992) Managing insect resistance to *Bacillus thuringiensis* toxins. *Science* 258:1451-1455.

McPartlan, H.C. and Dale, P.J. (1994) An assessment of gene transfer by pollen from field-grown transgenic potatoes to non-transgenic potatoes and related species. *Transgenic Research* 3:216-225.

Mohammed, A., Douches, D.S., Pett, W., Grafius, E., Coombs, J., Liswidowati, L.W., and Madkour, M.A. (2000) Evaluation of potato tuber moth (*Lepidoptera: Gelechiidae*) resistance in tubers of *Bt-cry5* transgenic potato lines. *Journal of Economic Entomology* 93(2):472-476.

Naimov, S., Weemen-Hendriks, M., Dukiandjiev, S., and de Maagd, R.A. (2001) *Bacillus thuringiensis* delta-endotoxin Cry1 hybrid proteins with increased activity against the Colorado potato beetle. *Applied and Environmental Microbiology* 67(11):5328-5330.

Oerke, E.C., Dehne, H.W., Schonbeck, F., and Weber, A. (1995) Crop production and crop protection: Estimated losses in major food and cash crops. En. Amsterdam (The Netherlands). Elsevier Science. 808 pp.

Ooms, G., Burrell, M.M., Karp, A., Bevan, M., and Hille, J. (1987) Genetic transformation in two potato cultivars with T-DNA from disarmed *Agrobacterium. Theoretical and Applied Genetics* 73:744-750.

Paarlberg, R. (2000) Governing the GM crop revolution: Policy choices for developing countries, pp. 251-255. *In*: Pinstrup-Andersen, P., and Pandya-Lorch, R. (eds.), The Unfinished Agenda: Perspectives on Overcoming Hunger, Poverty, and Environmental Degradation. International Food Policy Research Institute, Washington DC, USA.

Peferoen, M., Jansens, S., Reynaerts, A., and Leemans, J. (1990) Potato plants with engineered resistance against insect attack, pp.193-204. *In*: Vayda, M., and Park, W. (eds.), Molecular and Cellular Biology of the Potato, CAB, Tucson, AZ, USA.

Perlak, F.J., Stone, T.B., Muskopf, Y.M., Petersen, L.J., Parker, G.B., McPherson, S.A., Wyman, J., Love, S., Reed, G., Biever, D., and Fischhoff, D.A. (1993) Genetically improved potato: protection from damage by Colorado potato beetles. *Plant Molecular Biology* 22:313-321.

Qaim, M. (1998) Transgenic virus resistant potatoes in Mexico: potential socioeconomic implications of North-South biotechnology transfer. *ISAAA* Briefs No. 7. International Service for the Acquisition of Agri-biotech Applications (ISAAA), Ithaca, NY, USA.

Quiros, C.F., Ortega, R., Van Raamsdonk, L., Herrera-Montaya, M., Cisneros, P., Schmidt, E., and Brush, S.B. (1992) Increase of potato genetic resources in their center of diversity: the role of natural outcrossing and selection by the Andean farmer. *Genetic Resources and Crop Evolution* 39:107-113.

Rabinowitz, D., Linder, C.R., Ortega, R., Begazo, D., Murguia, H., Douches, D.S., and Quiros, C.F. (1990) High levels of interspecific hybridization between *Solanum sparsipilum* and *S. stenotomum* in experimental plots in the Andes. *American Potato Journal* 67:73-81.

Radcliffe, E.B. and Lagnaoui, A. (1990) Potential of neem for control of pyrethroid-resistant Colorado potato beetle, pp. 57-66. *In*: Neem's Potential in Pest Management Programs USDA Neem. Proceedings of a Workshop, Beltsville Agricultural Research Center, Beltsville, Maryland 16-17 April, 1990. ARS-86.

Raman, K.V. (1994) Potato pest management in developing countries, pp. 583-598. *In*: W. Zehnder, M. Powelson, R. Jansson, K Raman (eds.). Advances in Potato Pest Biology and Management. APS Press. 655 pp.

Roux, O., Von Arx, R., and Baumgärtner, J. (1992) Estimating potato tuberworm (Lepidoptera: Gelechiidae) damage in stored potatoes in Tunisia. *Journal of Economic Entomology* 85:2246-2250.

Saxena, A.P. and Rizvi, S.M.A. (1974) Insect pest problems of potato in India. *Journal of Indian Potato Association* 1:45-30.

Schlüter, K., Fütterer, J., and Potrykus, I. (1995) "Horizontal" gene transfer from a transgenic potato line to a bacterial pathogen (*Erwinia chrysanthemi*) occurs–if at all–at an extremely low frequency. *Biotechnology* 13:1094-1098.

Skogsmyr, I. (1994) Gene dispersal from transgenic potatoes to conspecifics: a field trial. *Theoretical and Applied Genetics* 88:770-774.

Spooner, D.M. and Hijmans, R. (2001) Potato systematics and germplasm collecting, 1989-2000. *American Journal of Potato Research* 78: 237-268; 395.

Spooner, D.M., Salas, A., Huamán, Z., and Hijmans, R. (1999) Wild potato collecting expedition in Southern Peru (Departments of Apurímac, Arequipa, Cusco, Moquegua, Puno, Tacna) in 1998: Taxonomy and new genetic resources. *American Journal of Potato Research* 76:103-119.

Stockhaus, J., Eckes, P., Blau, A., Schell, A., and Willmitzer, L. (1987) Organ-specific and dosage-dependent expression of a leaf/stem specific gene from potato after tagging and transfer into potato and tobacco plants. *Nucleic Acid Research* 15(8): 3479-3491.

Tabashnik, B.E. (1994) Evolution of resistance to *Bacillus thuringiensis*. *Annual Review of Entomology* 39:47-79.

Tamm, L. (1999) Current situation of organic potato production in Europe. Proceedings from the GILB Conference 16-19 March 1999, Quito, Ecuador. *Late blight: A threat to global food security.* p. 92.

Traxler, G., Godoy-Avila, S., Falck-Zepeda, J., and Espinoza-Arellano, J. (2001) Transgenic Cotton in Mexico: Economic and Environmental Impacts. Electronic Proceedings of the 5th International Conference on Biotechnology, Science and Modern Agriculture. International Consortium on Agricultural Biotechnology Research (ICABR) held in Ravello, Italy. June 15-18, 2001 (*http://www.economia. uniroma2.it/conferenze/icabr01/Program.htm*).

Van Rie, J., Jansens, S., and Reynaerts, A. (1994) Engineered resistance against potato tuber moth, pp. 499-508. *In:* Zehnder, G.M., Powelson, M.L., Jansson, R.K., and Raman, K.V. (eds.), Advances in Potato Pest Biology and Management, The American Phytopathological Society press.

Walker, T.S., and Collion, M.-H. (1997) Priority setting at CIP for the 1998-2000 Medium Term Plan. International Potato Center, Lima, Peru. 58 pp.

White, J.W. (1983) Pollination of potatoes under natural conditions. *CIP Circular* 11(2):1-2. International Potato Center (Peru).

Winters, P. and Fano, H. (1997) The economics of biological control in Peruvian potato production. Social Science Department, Working paper No. 1997-7. International Potato Center (CIP), Lima, Peru.

Ecological Impact of Bt Cotton

Allan E. Zipf
Kanniah Rajasekaran

SUMMARY. An overview of the ecological impact of transgenic crops, especially Bt cotton, is given in this paper. Crops expressing insecticidal crystal protein (ICP) genes from *Bacillus thuringiensis* (Bt) were among the first transgenic products approved for commercial use in the USA and several other countries. The cotton farming community in the USA embraced the transgenic technology because of potential benefits such as reducing (1) costs of operation and (2) damage to the environment and ground water supply, due to the repeated use of pesticides. However, concerns about genetically modified crops and foods remain in the USA and elsewhere. The transfer of Bt genes to wild relatives and neighboring crops, impact of Bt on non-target organisms, effects on crop yield and the possibility of insect pest populations developing resistance have all been items of concern. Key aspects of these benefits and concerns are summarized here. *[Article copies available for a fee from The Haworth Document Delivery Service: 1-800-HAWORTH. E-mail address: <docdelivery@ haworthpress.com> Website: <http://www.HaworthPress.com> © 2003 by The Haworth Press, Inc. All rights reserved.]*

Allan E. Zipf is Research Associate Professor, Department of Plant and Soil Science, Alabama A&M University, Normal, AL 35762.

Kanniah Rajasekaran is Research Biologist, USDA, ARS, Southern Regional Research Center, New Orleans, LA 70124.

Address correspondence to: Kanniah Rajasekaran, USDA/ARS, Southern Regional Research Center, 1100 Robert E. Lee Boulevard, New Orleans, LA 70124 (E-mail: krajah@srrc.ars.usda.gov).

The authors appreciate the valuable suggestions for improvement by Tony DeLucca and Jay Mellon.

[Haworth co-indexing entry note]: "Ecological Impact of Bt Cotton." Zipf, Allan E., and Kanniah Rajasekaran. Co-published simultaneously in *Journal of New Seeds* (Food Products Press, an imprint of The Haworth Press, Inc.) Vol. 5, No. 2/3, 2003, pp. 115-135; and: *Bacillus thuringiensis: A Cornerstone of Modern Agriculture* (ed: Matthew Metz) Food Products Press, an imprint of The Haworth Press, Inc., 2003, pp. 115-135. Single or multiple copies of this article are available for a fee from The Haworth Document Delivery Service [1-800-HAWORTH, 9:00 a.m. - 5:00 p.m. (EST). E-mail address: docdelivery@haworthpress. com].

http://www.haworthpress.com/store/product.asp?sku=J153
© 2003 by Taylor & Francis.
10.1300/J153v05n02_03

KEYWORDS. Biotechnology, Bt, cotton, environmental impact, ecology, insect resistance, gene flow, resistance management, transgenic crops, pesticides

Crop biotechnology, the science of moving gene(s) across sexual barriers, is viewed as an extension of traditional cross-breeding. The recent controversy over acceptance of genetically modified crops has created a substantial barrier to development and utilization of a technology with much promise for mankind (Rajasekaran et al., 2002; Radin and Bretting, 2002). The purpose of this paper is to address potential environmental impacts of transgenic crops, taking Bt cotton as a case study. The science behind the production of transgenic Bt cottons was reviewed recently by Wilkins et al. (2000) and Rajasekaran et al. (2001). The economic, food safety, social, and ecological consequences of the deployment of Bt transgenic plants in general have recently been reviewed by Shelton et al. (2002).

ECOLOGICAL IMPACT OF Bt COTTON

Discussion of the ecologic impact of transgenic Bt cotton extends to genes other than single Bt genes because (1) there is more than one type of Bt ICP available, including stacked varieties (e.g., Bollgard II™, Cry1Ac + Cry2Ab Bt Cotton); (2) some cultivars offer Bt bundled with herbicide resistance (e.g., Bollgard/RRCotton™); and (3) there are accessory antibiotic resistance genes used for bacterial and/or callus selection during the transformation process. Therefore, previously voiced pros and cons with other transgenic crops over vertical and horizontal spread of antibiotic resistance genes, increased weediness/problems associated with herbicide-resistant seed carryover and gene flow (e.g., Marvier, 2001; Obrycki et al., 2001; Hails, 2000; Bergelson et al., 1999) are also relevant to Bt cotton.

Much of the information on ecological effects of the Bt transgene has been generated in response to transgenic corn cultivars, often resulting in a contentious debate in the popular press and on the Internet. Having said this, unfortunately, there is limited peer-reviewed information on the impact of Bt cotton on the environment in the USA. Some published reports on plants containing the Bt insecticidal gene can be easily extrapolated to Bt cotton while such a connection is not as straight forward with other studies.

The concerns over the ecology of Bt cotton have been grouped under several broad areas–gene flow, development of pest resistance, pesticide use, impacts on non-target organisms, and yield and quality of fiber and cottonseed–with the understanding that these areas overlap in their ramifications. For example, the reduction of insecticidal spraying with Bt cotton leads to an increase of insect species, both pests and predators. The population dynamics are complex and not currently understood. Is there an inhibitory effect on predators preying on pest insects that had fed on Bt cotton, thus altering the dynamics of predator-prey interactions? Similarly, with reduced sprays, do pollinators increase in number, leading to potentially increased pollen transfers (gene flow) between wild relatives or neighboring fields, leading to 'super weeds' or marketing nightmares (Thayer, 1999)?

GENE FLOW OR ESCAPE

Concerns about flow of transgenes from genetically engineered crops to other sexually compatible plants can be broken into three general areas: (1) disruption of genetic diversity in cultivation systems; (2) disturbance of ecosystems by creation of heartier weeds; and (3) mingling with crops intended for markets that reject genetically engineered crops resulting in economic losses. The possibility of transgenes moving from genetically engineered plants into other organisms across sexual barriers has also been imagined as a potential hazard, but there are no known special properties of transgenes that would facilitate this to a higher degree than what occurs with DNA in general.

Gene flow or escape became a subject of intense scientific debate following a recent publication claiming that transgenic DNA had become genetically incorporated into traditional landraces of maize in Mexico (Quist and Chapela, 2001; Christou, 2002). The presence of the transgenes in maize landraces is not disputed by scientists but the implications for possible negative impacts on either maize genetic diversity or on the environment are not resolved.

Although "weediness" of Bt cotton has not been directly addressed, the lack of anecdotal evidence on weedy Bt cottons since 1995 is suggestive, but not proof. Concerns over "weediness" of transgenic crops have somewhat been alleviated by Crawley et al. (2001) who found no increased fitness of 4 modified crops in 12 different habitats after 10 years. A recent study in sunflower indicated that transgenic disease protection did not confer a selective advantage when crossed into the wild relative (Burke and Riesberg, 2003). Obviously, these results do not give *carte blanche* to every transgenic crop now or in the future. However, development of "triple herbicide-resistant" canola (Hall et al., 2000), arising as volunteers due to pollen flow between herbicide-resistant *Brassica napus*, suggests the necessity of very careful planning of future transgenic cottons.

Though cotton is considered to be a self-pollinated crop, in the presence of pollinators, outcrossing can be anywhere from 10-90+% (McGregor, 1976; Oosterhuis and Jernstedt, 1999). The spread of Bt genes from cotton is intimately tied to pollinator activity as the pollen grains are coated with a viscid material that causes them to adhere to each other, making wind dispersal likely to be an insignificant factor (McGregor, 1976). In general, though honeybees actively forage in cotton fields, they are not thought to actively collect the relatively large cotton pollen with its spiny exine (e.g., Loper and Degrandi-Hoffman, 1994; Vaissiere and Bradleigh, 1994). Cotton pollen can sporadically be found on pollen loads from honeybees hived near cotton fields (A. Zipf, K. Ward and R. Ward, personal observation), but very little is transferred to workers to result in cross-pollination from in-hive pollen transfer (Loper and Degrandi-Hoffman, 1994). However, anecdotal evidence from beekeepers suggests foragers will actively collect cotton pollen, especially if no other flowers are close by. It may be that beehives will have to be placed in the middle of cotton fields to reduce the chances of unintended cross-pollination, similar to conditions suggested for alfalfa (*http://www.ars.usda.gov/is/AR/archive/oct01/pollen1001.htm*).

Pollinator activity, of course, is correlated with the frequency and types of insect control. Insect control in Bt cotton is obviously related to the types of pests and their numbers that are dependent on region and weather conditions. However, in general, the reduced number of sprays for insect control in Bt cotton has led to the return of numerous insects and even to the return of using honeybees for improved yield/seed set and cotton honey production (Ward and Ward, 2001).

Transfer of transgenes to wild species of cotton is felt to be of minimal risk in the USA. There are three wild relatives of Bt cotton in the USA–*Gossypium thurberi*, *G. tomentosum* and feral *G. hirsutum. G. thurberi*, the wild diploid relative found in Arizona, is not compatible with pollen from tetraploid *G. hirsutum. G. tomentosum* occurs on six islands in Hawaii and is considered reproductively isolated from *G. hirsutum* or *G. barbadense* (DeJoode and Wendel, 1992). Because of the uncertainty of reproductive or phenological isolation, sale or use of Bt cotton has been banned in Hawaii by the Environmental Protection

Agency (EPA). Wild *G. hirsutum* cottons are found only in parts of southern Florida, almost exclusively in the Everglades National Park and the Florida Keys, hence the EPA restriction on sale and distribution of Bt cotton to those sites north of Tampa (Route 60). Therefore, in the USA, the potential to cross with wild species of cotton is minimal. However, this does not address the concerns of pollen transfer between neighboring transgenic and non-transgenic fields or seed movement between neighboring countries (Mexico/USA).

Cotton pollen dispersal by pollinators dropped to less than 1% outcrossing 4-7 m away from test plants but this low level could be sporadically detected up to 25 m away from the test plants in Australia and the USA (Llewellyn and Fitt, 1996; Umbeck et al., 1991). The distance of 25 m was considered to provide adequate containment for transgenic varieties (Umbeck et al., 1991) but that was before intense controversy over genetic engineering of crops.

Ultimately, the degree of gene flow will depend on pollinator activity in cotton. Bumblebees (*Bombus* spp.), Melissodes bees, and honeybees (*Apis mellifera*) are the primary pollinators of Upland cotton, each with different foraging ranges and fidelity. Pre-WWII cotton breeders suggested separations of 1-10 miles to insure isolation. The isolation distance for Foundation, Registered, and Certified seed in 7 CFR Part 201 is 1320 feet, 1320 feet, and 660 feet, respectively.

However, the current zero tolerance for any GMO traits in some markets and the incredible sensitivity of the detection systems may result in a reconsideration of these distances as the producer may encounter possible dockages or restricted markets. Although little concern currently exists regarding how much mixing of pollen occurs when two "conventional" varieties are planted next to each other (whether detection could even be possible), the current technology can theoretically detect single transgenic pollen transfers, thus "tainting" an entire harvest (Spiegelhalter et al., 2001).

A paradox of the debate over genetic engineering is that the so-called "Terminator Technology" could offer a solution to eliminating just such a gene transfer, but the technology is opposed by those with strong concerns about transgenic plants. This technology creates a sterile, but otherwise productive plant, such that viable offspring, and the resulting gene flow concerns are eliminated (Oliver et al., 1998). This technology has the dual outcome of both preventing gene flow, but also necessitating that farmers continually purchase seed, because the crop does not make viable seed. Cotton is not often sold as a hybrid seed, and is thus a likely candidate for terminator technology protection. By way of contrast, corn is usually planted as a hybrid, and thus has some measure of built-in variety protection already.

Another emerging technology that is very promising to contain the gene flow or escape via pollen spread is plastid transformation (as compared to nuclear transformation). Plastid (e.g., chloroplast) transformation has the advantage of yielding very high gene expression, as well as maternal inheritance in most of the crop species. The pattern of plastid inheritance alleviates concerns over gene flow via pollen since plastids, and any transgenes in them, are excluded from the pollen (Maliga, 2002; Daniell, 2002).

Another gene flow scenario that has been proposed is the possibility of horizontal gene transfer from transgenic plants to soil-dwelling or enteric bacteria. Though transfer has been demonstrated under laboratory conditions, it is still unclear if it happens in the field (Nielsen et al., 1998). The most likely instance would be exchange through homologous recombination of bacterial-derived transgenes, but the impact of exchanging such eukaryotic/prokaryotic gene fusions is little known (Kay et al., 2002). The prospect of bacteria gaining antibiotic resistance from transgenic plant material is insignificant compared to the availability of antibiotic resistance genes from other bacteria prevalent under natural conditions (Nester et al., 2002).

Limited information is available on possible effects of transgenes in animal systems. Though fragments of non-transgenic plant genes could be detected in the intestinal tract up to 121 hours after digestion and in the liver and spleen of mice fed soybean leaves, there was no indication of their expression, based on RT-PCR. Mice fed daily with GFP DNA for over 8 generations also showed no indication of germline transfer of ingested DNA (Hohlweg and Doerfler, 2001).

NON-TARGET ORGANISMS

That Bt corn pollen affected Monarch butterfly larvae was not surprising; after all, the active ingredient, Cry1Ac is active against Lepidoptera. Studies of Bt corn events suggested lethality was dose-dependent for Black (Wraight et al., 2000; Zangerl et al., 2001) and Eastern Tiger and Spicebush (Scriber, 2001) Swallowtail and Monarch (Sears et al., 2001) butterflies, but the likelihood of encountering lethal doses in the cornfield was very low. A detailed review of the interactions of Bt with non-target organisms is give by B. Federici (this volume, pp. 11-13).

Feeding of Bt toxin/Bt pollen/Bt plants to non-target organisms has reinforced the selective nature of the toxin. Concern, however, has been raised over reports of reduced viability/fecundity of predators after feeding on Bt intoxicated prey (e.g., Hilbeck et al., 1998). Additional predator/prey studies that report no (e.g., Head et al., 2001; Escher et al., 2000) or even positive (e.g., Schuler et al., 1999) effects cloud the issue leading to speculation that inadequate diet may be responsible for the reduction and not a toxic effect. There was also a difference in Bt uptake by phytophagous insects between artificial diets containing Bt toxins vs. feeding on transgenic Bt corn plants (Head et al., 2001). However, further study is clearly needed, especially on those predators/parasites that impact cotton pests.

As with any agroecosystem, disturbing one factor changes the responses of the other components. Use of Bt cotton in controlling tobacco budworm has led to reported increases in other secondary pests (e.g., fall armyworm), requiring increased insecticidal treatments (Hardee et al., 2001; Roberts, 1999). Concomitantly, there have also been increases in numerous beneficial insects (Roberts, 1999; Roof and Durant, 1997), forcing a revision of the integrated pest management (IPM) practices for Bt cotton to balance the pros and cons.

Bt cotton has been reported to have minimal or no effect on piercing/sucking predators (Armstrong et al., 2000) or on honeybees, ladybugs, spiders, big-eyed and pirate bugs and parasitic wasps (Hardee et al., 2001). Transgenic Bt cotton toxin sources (2 events) had no negative effects on egg production of 2 nontarget soil arthropods, a collembolan, *Folsomia candida* Willem, and an orbatid mite, *Oppia nitens* Koch (Yu et al., 1997). A second study of Bt cotton also indicated no significant effects on two collembolans, *F. candida* and *Xenylla grisea* (Sims and Martin, 1997).

Possible effects of Bt cotton on the rhizosphere microflora (species composition, population levels and carbon content) have been addressed by Donegan et al. (1995) and Donegan and Seidler (1998). Under field conditions, the rhizosphere microflora associated with Bt cotton leaves placed into soil differed minimally or were transiently increased from that of incorporated conventional cotton leaves or pure ICP treatments monitored 28 or 56 days later. It was thought that metabolic alteration of the plants during culturing, and not from Bt gene expression, was responsible for the transient effects (Donegan et al., 1995).

Two separate transgenic events of Bt cotton released measurable quantities, proportional to the original levels, of the ICP when incorporated into soil and immediately extracted (Palm et al., 1994). Extractable Bt cotton ICP was either undetectable or over 1/3 still present, starting from 1 or 1600 ng Bt ICP (leaf)/g soil, respectively, after 140 days (Palm et al., 1996). However, pure Bt ICP was also still detectable over the same time frame when added to the same soil. Less than 25% bioactivity remained from buried Bt cotton or pure toxin after 120 days burial over the fall and winter in MO (Sims and Ream, 1997).

Retention in the soil is obviously related to soil makeup, with increased clay (esp. montmorillonite and kaolinite) resulting in longer lifetimes (Tapp et al., 1994). However, increased clay also decreases the recovery of the ICP as well (Palm et al., 1994). Release of Bt ICP into the rhizosphere and its reduced degradation has also been documented from Bt corn and Bt sugar beets (e.g., Saxena et al., 1999; Saxena and Stotzky, 2000; Gebhard and Smalla, 1999). However, the lack of effect from these exudates on non-target organisms has also been reported (e.g., Saxena and Stotzky, 2001). Reports of rapid degradation of Bt ICP within soil and within leaves buried within soil (see Sims and Holden, 1996) illustrate the variability within the agrisphere, reducing the opportunity for blanket conclusions. Though many lepidopterans pupate in the soil, there are very few that feed in the soil. Therefore, effects on lepidopterans with a soil phase, from Bt corn or Bt cotton ICPs leaching from roots are thought to be minimal, if any.

Land use will thus play a large role in Bt ICP retention, depending on single/double/triple cropping and rotation systems. Bt cotton followed by a non-Bt crop or fallow would allow for degradation in the soil before the next Bt crop. However, Bt cotton followed by Bt corn, for example, may never allow complete degradation of Bt ICP, again depending on rhizosphere dynamics.

PESTICIDE USE

The impact on farming practices of Bt cotton vs. non-Bt cotton are relative, depending upon weather conditions, topography/region and year-to-year insect pressure(s). Therefore, there is a wide gamut of benefits/detriments, depending on year and region (summarized in Marra et al., 2002; Edge et al., 2001). For example, insect control costs were less than half with the transgenic Bt cotton than with non-Bt cotton in the USA (Demaske, 1997). Edge et al. (2001) evaluated 12 studies from 1995-1999 from across the world and found the average number of spray reductions in each study ranged from 1 to almost 8, with an overall average reduction of 3.5. In the US, reductions in number of pesticide treatments in six cotton-growing states have been reported (Ginanessi and Carpenter, 1999). According to this report, an average of 1.4 spray treatments were used for insect-protected cotton during 1996-98, compared to an average of 5.3 spray treatments for conventional cotton. This reduction in pesticide usage eliminated the use of approximately 2.0 million pounds of insecticides. Similarly, total insecticide use in China was reduced by 60-80% (Xia et al., 1999; Huang et al., 2002). Elena de Bianconi describes universal reductions in insecticide use in a two-year study of cotton grower practices in Argentina (this volume, pp. 223-235). In general, adoption of Bt cotton has led to reduced pesticide spraying, reduced use of conventional pesticides, increased use of biological/alternative pesticides and increased use of beneficial organisms (Marra et al., 2002).

Reduced pesticide use associated with Bt cotton may also result in reduced insecticide runoff, though only one study in MS (Cullum and Smith, 2001) has been conducted so far. An interesting observation has been the increased numbers of birds counted in cotton fields since the adoption of Bt cotton (EPA, 2001), though exact cause and effect cannot be proven at this time.

On the other hand, reduced spraying for heliothine insects has sometimes been accompanied by increased spraying for other pests, again depending on region and year (e.g., Layton et al., 2000). And there have also been reports of reduced applications for non-Bt-targeted pests as well (Benedict and Altman, 2001).

YIELD AND FIBER QUALITY

The primary ecological benefit of increased yield is a potential reduction in the demand for clearing of wild ecosystems for farmland. Yield of transgenic crops have either increased or decreased, depending on year, region, weather, or pest pressure (summarized in Edge et al., 2001). This variation is quite expected because predominant factors, in addition to pest pressure, that influence yield improvement are cultivar-specific qualities and climate variation (Meredith, 2002).

In general, however, yields for Bt cotton in the USA and worldwide have either equaled or exceeded yields of conventional varieties (e.g., Jenkins et al., 1997; Mahaffey et al., 2000; Robinson and McCall, 2001; Moser et al., 2001; Meredith, 2002; Huang et al., 2002; Qaim and Zilberman, 2003). Whether these offset the increased seed costs also varies from year to year, region to region, even within a state (e.g., Bryant et al. 2002). However, economic benefits of Bt cotton use by smallholder farmers have been documented in several countries. For example, according to Ismael et al. (2002), Bt growers in South Africa realized higher revenues than non-Bt growers, primarily due to higher yields and reduced pesticide costs. Similar benefits to small farmers in China and Mexico have been highlighted in studies by Pray et al. (2001) and Traxler et al. (2001), respectively, as well as in Argentina by Elena de Bianconi (this volume, pp. 223-235).

Similar to yield, fluctuations in fiber quality have been reported from several cotton growing states within the USA. Following the adoption of transgenic Bt and Roundup Ready cotton varieties, it was speculated that some areas of the US have had lower than normal fiber quality of transgenic varieties compared to their recurrent parents (Kerby et al., 2000). These authors compared fiber quality (staple length, fiber strength, and micronaire) of seven Deltapine varieties and their transgenic versions and concluded that the transgenic varieties were equivalent to their conventional parents. In another evaluation of transgenic (Bt and Roundup Ready) and parent varieties for fiber and spinning performance, no statistically significant differences were detected in yarn or fiber quality (Ethridge and Hequet, 2000). In contrast, Blanche et al. (2002) reported that even though transgenic plants grew taller and had heavier seed, conventional parents had higher lint percents and all cultivars yielded equally over four environments in Louisiana.

SEED QUALITY

Cottonseed products are used as food and feedstuffs and, thus, must be tested for safety. GM crops, in general, have been judged to be nutritionally the same as their non-GM counterparts; however, the only public information for Bt cotton is on proximate analysis (crude protein and fat, ash, amino acid, fatty acid and moisture levels) and antinutrient levels of cottonseed for animal feeds (Berberich et al., 1996).

Sims et al. (1996) determined the concentration of transgenic protein in insect-protected or glyphosate-resistant varieties. They concluded that typical cotton fiber processing steps prior to food or textile use will reduce transgenic proteins to undetectable levels. Similar analyses of glyphosate-tolerant cotton products demonstrated that they were compositionally and nutritionally comparable to and as safe as conventional cotton varieties (Nida et al., 1996).

Another positive, albeit indirect, effect of Bt cotton on improved safety of animal feed from cottonseed is worthy of mention. A primary entry route for the aflatoxin-producing fungus, *Aspergillus flavus*, into the cottonseed is through pink bollworm exit holes in unopened bolls. It has been shown that mycotoxin contamination was greatly reduced in Bt corn varieties due to reduction in insect damage (Munkvold et al., 1998, 1999; Dowd, 2000). By logical extension, preharvest aflatoxin contamination in Bt cotton would thus be greatly reduced due to minimization of exit holes caused by pink bollworms. Unfortunately, producers often leave harvested cotton (Bt or non-Bt) in storage modules in the field for an extended period after harvest, subjecting it to considerable weathering and allowing for post-harvest aflatoxin contamination (Cotty et al., 1997). Thus, prudent management practices are necessary for Bt cotton, as well as non-Bt cotton, in order to keep aflatoxin contamination to a minimum.

DEVELOPMENT OF Bt RESISTANT PESTS

Insect pests developing resistance to Bt is probably the most frequently voiced concern about Bt crops, not only from the organic farming sector, which relies heavily on Bt-related products, but also from ecologists and producers, concerned about "super pests" developing from the continuous use of Bt crops. This issue is also especially important to those who value the utility of Bt crops, and desire to see their usefulness maintained.

If development of Bt resistance is analogous to that of other pesticides, then resistance development will vary from species to species and from Bt type to Bt type, but cannot be dismissed. The current US regulatory requirements for management to counter development of Bt resistant pests, a high dose/refugia model, is treated in detail by Matten and Reynolds (this volume, pp. 137-178). Variations in toxin levels between plant organs (buds vs. bolls vs. anthers, etc.) as well as differing expression levels due to environment and/or promoter interactions (e.g., Greenplate, 1999) or even plant physiology-ICP interactions (Olsen and Daly, 2000) complicate effective, consistent resistance management strategies that rely on a high dose. A major assumption in all models was that all producers "play by the rules," a circumstance that may not be true for some/all crops (Dove, 2001).

In studies with resistant insect species selected against pure toxins or spore formulations, the resistance extended to more than one Bt ICP (reviewed in Ferre and Rie, 2002). This strong cross-resistance occurred only with ICPs apparently sharing insect gut binding sites and not from within the same Cry families (i.e., little sequence similarities). Cross-resistance has major implications for choice of pyramided Bt toxins in future transgenic crop cultivars. Unfortunately, binding site specificities apparently differ between insects (e.g., *Plodia xylostella* [diamondback moth] vs. *Heliothis virescens* [tobacco budworm]), further complicating future toxin combination choices.

Of specific relevance to Bt cotton, resistance in *H. virescens* colonies reared on Cry1Ac protoxin cross-reacted to several other Bt toxins, including Cry1Ab, Cry1Fa (Gould et al., 1995), Cry2Aa, Cry1Aa, Cry1Ba and Cry1Ca (Gould et al., 1992).

The resistance to Bt toxins was most often due to alteration of toxin binding to the gut wall, although other mechanisms included protease degradation of protoxins and improved repair of the damaged midgut cells, indicating other loci for resistance (Ferre and Rie, 2002). Although the resistance often reverted under non-selection, that was not always the case with these laboratory studies. However, these resistant/tolerant individuals may be more susceptible to follow-up sprays of conventional pesticides (Harris et al., 1998). Both multiple resistance genes and resistance mechanisms have been found in the cotton pest, *H. virescens* (reviewed in Ferre and Rie, 2002; Adamczyk and Hardee, 2002; Carpenter et al., 2002; Walker et al., 2002).

The initial refugia recommendations were made based on the models that assumed that resistance alleles/loci were rare, recessive, and single. For studies performed with Bt crops, resistance was shown to be completely recessive. However, resistance developing in the laboratory in response to Bt formulations or pure proteins ranged from partially dominant to partially recessive to completely recessive (reviewed in Ferre and Rie, 2002). If such resistance develops in the field, significant changes in resistance management strategies will be needed.

Of most concern were the findings that resistance gene frequencies were not as low as expected, possibly compromising the existing resistance management strategy of refugia (Ferre and Rie, 2002). In addition, variations in susceptibility to Cry1Ac, over 12-fold, have been found in *Heliothis virescens* (Stone and Sims, 1993) and *Helicoverpa zea* (Luttrell et al., 1999; Stone and Sims, 1993) populations.

Tolerance to Cry1Ac already appeared by 1996 (one year after release of Bollgard cotton) in populations of *Helicoverpa zea*, but not *Heliothis virescens*, sampled from across the southeastern USA (Sumerford et al., 1999). Levels of tolerance had increased in *H. zea* populations by

1999, associated with the amount of Bt cotton grown. A single *H. virescens* population from Mississippi had significant Bt tolerance by then as well. However, sampling of populations of pink bollworm, *Pectinophora gossypiella*, showed no increase in resistance gene frequency after two subsequent seasons despite the selection pressure of Bt cotton (Tabashnik et al., 2000).

Adding to the complexity of managing resistance development, any resistance management strategy should include behavioral factors, such as the migratory/flight ability of the targeted insect(s). For example, *Heliothis virescens* larvae dispersed more frequently from Bt cotton plants than from untransformed controls (Parker and Luttrell, 1999). However, neither oviposition, vertical placement of eggs or plant site selection differed for *H. virescens* females on Bt or conventional cottons (Parker and Luttrell, 1998).

Refuge distance also plays a role in resistance management. Most male pink bollworms move less than 400 m, indicating that any refugia need to be close to the Bt cotton for effective mating of resistant and susceptible individuals (Tabashnik et al., 1999). Thus, the placement of external refuge(s) should be determined by the dispersal habits of the targeted pest(s) (Caprio, 1998).

Complicating cotton pest controls are Bt cotton impacts on sensitive, yet non-registered targets (e.g., pink bollworm, *Pectinophora gossypiella*). Pink bollworm is less sensitive to Bollgard™; therefore, there is an opportunity for more numbers to survive, etc., to possibly build a resistant population (given the cross-reactivity of Bt resistance) before a Bt cotton specific to this pest is introduced. Bt cotton is also not fatal to fall armyworm, causing only delayed larval development (Adamczyk and Sumerford, 2001). After a single generation, offspring of fall armyworm parents who survived feeding on Bt cotton, though of significantly lower weight, had no significant differences in fitness, vigor or capability to reproduce (Adamczyk and Sumerford, 2001). Similar reductions in susceptibility were found in soybean loopers, *Pseudoplusia includens* Walker, collected from Bt cotton (Mascarenhas et al., 1998). The insecticidal spectrum of Bollgard II includes pink bollworms and beet and fall armyworms (Perlak et al., 2001).

In addition, there are the major problems of (1) what effect(s) will insects that move from crop to crop as the season progresses have on resistance management (e.g., cotton bollworm, *Helicoverpa zea* [corn → cotton] and fall armyworm, *Spodoptera frugida* [corn + cotton]) and (2) what effect(s) will different Bt crops grown in the same region have on resistance management strategies?

CONCLUSION

Bt cotton has the clearly demonstrated capacity to increase yields and reduce inputs of synthetic chemical insecticides. Some hazards ascribed to Bt and other genetically engineered crops, such as gene transfer to microbes, are more or less irrelevant. Other concerns, such as gene flow to wild relatives can be of particular concern depending on the potential for interbreeding, and the transgene in question; Bt and other insect protectant traits may have particular potential to perturb certain ecosystems. Realizing the full potential of Bt cotton, other Bt crops, and genetically engineered crops in general will require management of potential ecological and economic hazards, and management to maintain the utility of genetically engineered traits.

REFERENCES

Adamczyk, J.J., Jr. and D.D. Hardee. (2002) Insect-resistant transgenic crops. In: K. Rajasekaran, T.J. Jacks and J.W. Finley (eds.), Crop Biotechnology. ACS Symposium Series No. 829, American Chemical Society, Washington, DC.

Adamczyk, J.J., Jr. and D.V. Sumerford. (2001) Increased tolerance of fall armyworms (Lepidoptera: Noctuidae) to Cry 1Ac δ-endotoxin when fed transgenic *Bacillus thuringiensis* cotton: impact on the development of subsequent generations. *Florida Entomologist* 84: 1-6.

Armstrong, J.S., J. Leser and G. Kraemer. (2000) An inventory of the key predators of cotton pests on Bt and non-Bt cotton in west Texas. Proc. Beltwide Cotton Conferences 2000. National Cotton Council, Memphis, TN, pp. 1030-1033.

Benedict, J.H. and D.W. Altman. (2001) Commercialization of transgenic cotton expressing insecticidal crystal protein. In: J.J. Jenkins and S. Saha (eds.), Genetic Improvement of Cotton: Emerging Technologies. Science Publishers, Enfield, NH.

Berberich, S.A., J.E. Ream, T.L. Jackson, R. Wood, R. Stipanovic, P. Harvey, S. Patzer and R.L. Fuchs. (1996) Safety assessment of insect-protected cotton: The composition of insect-protected cottonseed is equivalent to conventional cottonseed. *J. Agric. Food Chem.* 44: 365-371.

Bergelson, J., J. Winterer and C.B. Purrington. (1999) Ecological impacts of transgenic crops. In: V.L. Chopra, V.S. Malik and S.R. Bhat (eds.), Applied Plant Biotechnology. Science Publishers, Enfield, NH.

Blanche, S.B., G.O. Myers, M. Akash and B. Jiang. (2002) Transgene effect on the stability of cotton cultivars in Louisiana. Proc. Beltwide Cotton Conferences 2002. National Cotton Council, Memphis, TN.

Bryant, K.J., W.C. Robertson, G.M. Lorenz, R. Ihrig and G. Hackman. (2002) Six years of transgenic cotton in Arkansas. Proc. Beltwide Cotton Conferences 2002. National Cotton Council, Memphis, TN.

Burke, J.M. and L.H. Riesberg. (2003) Fitness effects of transgenci disease resistance in sunflowers. *Science* 300:1250.

Caprio, M.A. (1998) Evaluating resistance management strategies for multiple toxins in the presence of external refuges. *J. Econ. Entomol.* 91: 1021-1031.

Carpenter, J., A. Felsot, T. Goode, M. Hammig, D. Onstad and S. Sankula. (2002) Comparative Environmental Impacts of Biotechnology-Derived and Traditional Soybean, Corn, and Cotton Crops. Council for Agricultural Science and Technology, Ames, IA. 190 pp.

Christou, P. (2002) No credible evidence is presented to support claims that transgenic DNA was introgressed into traditional landraces in Oaxaca, Mexico. *Transgenic Res.* 11: iii-v.

Cotty, P.J., D.R. Howell, C. Block and A. Tellez. (1997) Aflatoxin contamination of Bt cottonseed. 1997 Cotton Report, University of Arizona College of Agriculture Series P-108. pp. 435-439.

Crawley, M.J., S.L. Brown, R.S. Hails, D.D. Kohn and M. Rees. (2001) Biotechnology: Transgenic crops in natural habitats. *Nature* 409: 682-683.

Cullum, R. and Smith, S., Jr. (2001) Bt cotton in Mississippi Delta Management Systems Evaluation Area: Insecticides in runoff. 1996-1999. U.S. Department of Agriculture, Agricultural Research Service, Oxford, MS.

Daniell, H. (2002) Molecular strategies for gene containment in transgenic crops. *Bio/Technol.* 20: 581-586.

DeJoode, D.R. and J.F. Wendel. (1992) Genetic diversity and origin of the Hawaiian Islands cotton, *Gossypium tomentosum. Am. J. Bot.* 79: 1311-1319.

Demaske, C. (1997) Notes from a Bt seminar–what have we learned? *Cotton Grower* 33: 32-33.

Donegan, K.K., C.J. Palm, V.J. Fieland, L.A. Porteous, L.M. Ganio, D.L. Schaller, L.Q. Bucao and R.J. Seidler. (1995) Changes in levels, species and DNA fingerprints of soil microorganisms associated with cotton expressing the *Bacillus thuringiensis* var. *kurstaki* delta-endotoxin. *Appl. Soil Ecol.* 2: 111-124.

Donegan, K.K. and R.J. Seidler. (1998) Effect of transgenic cotton expressing the *Bacillus thuringiensis* var. *kurstaki* endotoxin on soil microorganisms–risk assessment studies. In Y.P.S. Bajaj (ed.) Biotechnology in Agriculture and Forestry, Vol. 42., Cotton. Springer, Berlin, pp. 299-312.

Dove, A. (2001) Survey raises concerns about Bt resistance management. *Nature Biotech.* 19: 293-294.

Dowd, P.F. (2000) Indirect reduction of ear molds and associated mycotoxins in *Bacillus thuringiensis* corn under controlled and open field conditions: Utility and limitations. *J. Econ. Entomol.* 93: 1669-1679.

Edge, J.M., J.H. Benedict, J.P. Carroll and H.K. Reding. (2001) Bollgard cotton: An assessment of global economic, environmental and social benefits. *J. Cotton Sci.* 5: 121-136.

EPA. (2001) Biopesticides Registration Action Document, Revised Risks and Benefits Sections: *Bacillus thuringiensis* Plant-Pesticides, July 16, 2001. U.S. Environmental Protection Agency, Office of Pesticide Programs, Biopesticides and Pollution Prevention Division, Washington, DC.

Escher, N., B. Kach and W. Nentwig. (2000) Decomposition of transgenic *Bacillus thuringiensis* maize by microorganisms and woodlice *Porcellio scaber* (Crustacea: Isopoda). *Basic Appl. Ecol.* 1: 161-169.

Ethridge, M.D. and E.F. Hequet. (2000) Fiber properties and textile performance of transgenic cotton versus parent varieties. *Proc. Beltwide Cotton Conferences 2000.* 1: 488-494.

Ferre, J. and J. van Rie. (2002) Biochemistry and genetics of insect resistance to *Bacillus thuringiensis. Annu. Rev. Entomol.* 47: 501-533.

Gebhard, F. and K. Smalla. (1999) Monitoring field releases of genetically modified sugar beets for persistence of transgenic plant DNA and horizontal gene transfer. *FEMS Microbiol. Ecol.* 28: 261-272.

Ginanessi, L.P. and J.E. Carpenter. (1999) Agricultural Biotechnology: Insect Control Benefits. Natl. Cent. Food Agric. Policy, Washington, DC.

Gould, F., A. Anderson, A. Reynolds, L. Bumgarner and W. Moar. (1995) Selection and genetic analysis of a *Heliothis virescens* (Lepidoptera: Noctuidae) strain with high levels of resistance to *Bacillus thuringiensis* toxins. *J. Econ. Entomol.* 88: 1545-1559.

Gould, F., A. Martinez-Ramirez, A. Anderson, J. Ferre, F.J. Silva and W.J. Moar. (1992) Broad-spectrum resistance to *Bacillus thuringiensis* toxins in *Heliothis virescens. Proc. Natl. Acad. Sci. USA* 89: 7986-7990.

Greenplate, J.T. (1999) Quantification of *Bacillus thuringiensis* insect control protein Cry 1Ac over time in Bollgard cotton fruit and terminals. *J. Econ. Entomol.* 92: 1377-1383.

Hails, R.S. (2000) Genetically modified plants–the debate continues. *Trends Ecol. Evol.* 15: 14-18.

Hall, L., K. Topinka, J. Huffman, L. Davis and A. Good. (2000) Pollen flow between herbicide-resistant *Brassica napus* is the cause of multiple-resistant *B. napus* volunteers. *Weed Sci.* 48: 688-694.

Hardee, D.D., J.W. van Duyn, M.B. Layton and R.D. Bagwell. (2001) Bt cotton and management of the tobacco budworm-bollworm complex. USDA, Agricultural Research Service, ARS-154. 40 pp.

Harris, J.G., C.N. Hershey and M.J. Watkins. (1998) The usage of Karate (l-cyhalothrin) oversprays in combination with refugia, as a viable and sustainable resistance management strategy for Bt cotton. Proc. Beltwide Cotton Conferences 1998. National Cotton Council, Memphis, TN, pp. 1217-1220.

Head, G., C.R. Brown, M.E. Groth and J.J. Duan. (2001) Cry1Ab protein levels in phytophagous insects feeding on transgenic corn: Implications for secondary risk assessment. *Entomol. Exp. Appl.* 99: 37-45.

Hilbeck, A., M. Baumgartner, P.M. Freid and F. Bigler. (1998) Effects of *Bacillus thuringiensis* corn-fed prey on mortality and development time of immature *Chrysoperia carnae. Eviron. Entomol.* 27: 480-487.

Hohlweg, U. and W. Doerfler. (2001) On the fate of plant or other foreign genes upon the uptake in food or after intramuscular injection in mice. *Mol. Genet. Genom.* 265: 225-233.

Huang, J, S. Rozelle, C. Pray and Q. Wang. (2002) Plant biotechnology in China. *Science* 295: 674-676.

Ismael, Y., R. Bennett and S. Morse. (2002) Benefits from Bt cotton use by small-holder farmers in South Africa. *AgBioForum* 5: 1-5.

Jenkins, J.N., J. McCarty, R.E. Buehler, J. Kiser, C. Williams and T. Wofford. (1997) Resistance of cotton with δ-endotoxin genes from *Bacillus thuringiensis* var. *kurstaki* on selected Lepidopteran insects. *Agron. J.* 89: 768-780.

Kay, E., T.M. Vogel, F. Bertolla, R. Nalin and P. Simonet. (2002) In situ transfer of antibiotic resistance genes from transgenic (transplastomic) tobacco plants to bacteria. *Appl. Environ. Microbiol.* 68: 3345-3351.

Kerby, T., B. Hugie, K. Howard, M. Bates, J. Burgess and J. Mahaffey. 2000. Fiber quality comparisons among varieties for conventional, Bollgard and Roundup Ready versions. *Proc. Beltwide Cotton Conferences 2000.* 1: 484-488.

Layton, M.B., M.R. Williams and J.L. Long. (2000) Performance of Bt cotton in Mississippi. *Proc. Beltwide Cotton Conferences 2000.* National Cotton Council, Memphis, TN. 2:1037-1039.

Llewellyn, D. and G. Fitt. (1996) Pollen dispersal from two field trials of transgenic cotton in the Namoi Valley, Australia. *Mol. Breed.* 2: 157-166.

Loper, G.M. and G. Degrandi-Hoffman. (1994) Does in-hive pollen transfer by honey bees contribute to cross-pollination and seed set in hybrid cotton? *Apidologie* 25: 94-102.

Luttrell, R.G., L. Wan and K. Knighten. (1999) Variation in susceptibility of Noctuid (Lepidoptera) larvae attacking cotton and soybean to purified endotoxin proteins and commercial formulations of *Bacillus thuringiensis. J. Econ. Entomol.* 92: 21-32.

Maliga, P. (2002) Engineering the plastid genome of higher plants. *Curr. Opin. Biotechnol.* 5: 164-172.

Mahaffey, J.S., K.D. Howard, T.A. Kerby, J.C. Burgess, M. Casavechia and A. Coskrey. (2000) The agronomic performance of one Bollgard II™ donor variety. *Proc. Beltwide Cotton Conferences 2000.* 1: 495-496.

Marra, M.C., P.G. Pardey and J.M. Alston. (2002) The payoffs to agricultural biotechnology: An assessment of the evidence. EPTD Discussion Paper No. 87. Environment and Production Technology Division, International Food Policy Research Institute, Washington, DC.

Marvier, M. (2001) Ecology of transgenic plants. *Am. Scientist* 89: 160-167.

Mascarenhas, R.N., D.J. Boethel, B.R. Leonard, M.L. Boyd and C.G. Clemens. (1998) Resistance monitoring to *Bacillus thuringiensis* insecticides for soybean loopers (Lepidoptera: Noctuidae) collected from soybean and transgenic Bt-cotton. *J. Econ. Entomol.* 91: 1044-1050.

McGregor, S.E. (1976) Insect Pollination of Cultivated Crop Plants. USDA, Carl Hayden Bee Research Center, Tucson, AZ. *<http://gears.tucson.ars.ag.gov/book/>*.

Meredith, Jr., W.R. (2002) Factors that contribute to lack of genetic progress. Proc. Beltwide Cotton Conferences 2002. National Cotton Council, Memphis, TN.

Moser, H.S., W.B. McCloskey and J.C. Silvertooth. (2000) Performance of transgenic cotton varieties in Arizona. *Proc. Beltwide Cotton Conferences 2000.* 1:497-499.

Munkvold, G.P., R.L. Hellmich and L.G. Rice. (1999) Comparison of fumonisin concentrations in kernels of transgenic Bt maize hybrids and non-transgenic hybrids. *Plant Dis.* 83: 130-138.

Munkvold, G., H.M. Starh, A. Logrieco, A. Moretti and A. Ritieni. (1998) Occurrence of fusaproliferin and beauvericin in *Fusarium*-contaminated livestock feed in Iowa. *Appl. Environ. Microbiol.* 64: 3923-3926.

Nester, E., L. Thomashow, M. Metz and M. Gordon. (2002) 100 Years of *Bacillus thuringiensis*: A Critical Scientific Assessment. A Report from the American Academy of Microbiology, Washington, DC.

Nida, D.L., S. Patzer, P. Harvey, R. Stiponovic, R. Wood and R.L. Fuchs. (1996) Glyphosate-tolerant cotton: the composition of cottonseed is equivalent to that of conventional cottonseed. *J. Agric. Food Chem.* 44: 1967-1974.

Nielsen, K.M., A.M. Bones, K. Smalla and J.D. van Elsas. (1998) Horizontal gene transfer from transgenic plants to terrestrial bacteria–a rare event? *FEMS Microbiol. Rev.* 22: 79-103

Obrycki, J.J., J.E. Losey, O.R. Taylor and L.C.H. Jesse. (2001) Transgenic insecticidal corn: Beyond insecticidal toxicity to ecological complexity. *Bioscience* 51: 353-361.

Oliver, M.J., J.E. Quisenberry, N.L.G. Trolinder and D.L. Keim. (1998) Control of Plant Gene Expression. US Patent 5,723,765, issued March 3, 1998.

Olsen, K.M. and J.C. Daly. (2000) Plant-toxin interactions in transgenic Bt cotton and their effect on mortality of *Helicoverpa armigera* (Lepidoptera: Noctuidae). *J. Econ. Entomol.* 93: 1293-1299.

Oosterhuis, D.M. and J. Jernstedt. (1999) Morphology and anatomy of the cotton plant. In: W.C. Smith and J.T. Cothren (eds.), Cotton: History, Technology and Production. John Wiley and Sons, Inc., New York, NY.

Palm, C.J., D.L. Schaller, K.K. Donegan and R.J. Seidler. (1996) Persistence in soil of transgenic plant produced *Bacillus thuringiensis* var. *kurstaki* delta-endotoxin. *Can. J. Microbiol.* 42: 1258-1262.

Palm, C.J., K.K. Donegan, D. Harris and R.J. Seider. (1994) Quantitation in soil of *Bacillus thuringiensis* var. *kurstaki* delta-endotoxin from transgenic plants. *Mol. Ecol.* 3: 145-151.

Parker, C.D.J. and R.G. Luttrell. (1998) Oviposition of tobacco budworm (Lepidoptera: Noctuidae) in mixed plantings of non-transgenic and transgenic cottons expressing delta-endotoxin protein of *Bacillus thuringiensis* Berliner. *Southwestern Entomologist* 23: 247-257.

Parker, C.D.J. and R.G. Luttrell. (1999) Interplant movement of *Heliothis virescens* (Lepidoptera: Noctuidae) larvae in pure and mixed plantings of cotton with and without expression of the Cry 1Ac delta-endotoxin protein of *Bacillus thuringiensis* Berliner. *J. Econ. Entomol.* 92: 837-845.

Perlak, F.J., M. Oppenhuizen, K. Gustafson, R. Voth, S. Sivasupramaniam, D. Heering, B. Carey, R.A. Ihrig and J.K. Roberts. (2001) Development and commercial use of Bollgard® cotton in the USA–early promises versus today's reality. *Plant J.* 27: 489-501.

Pray, C.E., D. Ma, J. Huang and F. Qiao. (2001) Impact of Bt cotton in China. *World Development* 29: 813-825.

Qaim, M. and D. Zilberman. (2003) Yield effects of genetically modified crops in developing countries. *Science* 299: 900-902.

Quist, D. and I.H. Chapela. (2001) Transgenic DNA introgressed into traditional maize landraces in Oaxaca, Mexico. *Nature* 414: 541-543.

Radin, J.W. and P.W. Bretting. (2002) Defining biotechnology: Increasingly important and increasingly difficult. In: K. Rajasekaran, T.J. Jacks and J.W. Finley (eds.), Crop Biotechnology. ACS Symposium Series No. 829, American Chemical Society, Washington, DC.

Rajasekaran, K., C.A. Chlan and T.E. Cleveland. (2001) Tissue culture and genetic transformation of cotton. In: J.J. Jenkins and S. Saha (eds.), Genetic Improvement of Cotton: Emerging Technologies. Science Publishers, Enfield, NH.

Rajasekaran, K., T.J. Jacks and J.W. Finley (eds.). (2002) Crop Biotechnology, ACS Symposium Series No. 829, American Chemical Society, Washington, DC. 260 pp.

Roberts, P. (1999) Observations of emerging pests in low spray environments. Proc. Beltwide Cotton Conferences 1999. National Cotton Council, Memphis, TN, p. 1034.

Robinson, M. and L. McCall. (2001) A comparison of transgenic and conventional cotton varieties. *Proc. Beltwide Cotton Conferences 2001.* 1: 419.

Roof, M.E. and J.A. DuRant. (1997) On-farm experiences with Bt cotton in South Carolina. Proc. Beltwide Cotton Conferences 1997. National Cotton Council, Memphis, TN, p. 861.

Saxena, D. and G. Stotzky. (2000) Insecticidal toxin from *Bacillus thuringiensis* is released from roots of transgenic Bt corn *in vitro* and *in situ. FEMS Microbiol. Ecol.* 33: 35-39.

Saxena, D. and G. Stotzky. (2001) *Bacillus thuringiensis* (Bt) toxin released from root exudates and biomass of Bt corn has no apparent effect on earthworms, nematodes, protozoa, bacteria, and fungi in soil. *Soil Biol. Biochem.* 33: 1225-1230.

Saxena, D., S. Elores and G. Stotzky. (1999) Insecticidal toxin in root exudates from Bt corn. *Nature* 402: 480.

Schuler, T.H., R.P.J. Potting, I. Denholm and G.M. Poppy. (1999) Parasitoid behavior and *Bt* plants. *Nature* 400: 825-826.

Scriber, J.M. (2001) *Bt* or not *Bt*: Is that the question? *Proc. Natl. Acad. Sci. USA* 98: 12328-12330.

Sears, M.K., R.L. Hellmich, D.E. Stanley-Horn, K.S. Oberhauser, J.M. Pleasants, H.R. Mattila, S.D. Siegfried and G.P. Dively. (2001) Impact of Bt corn pollen on monarch butterfly populations: a risk assessment. *Proceedings of the National Academy of Sciences U.S.A.* 98: 11937-11942.

Shelton, A.M., J.-Z. Zhao and R.T. Roush. (2002) Economic, ecological, food safety, and social consequences of the deployment of Bt transgenic plants. *Annu. Rev. Entomol.* 47: 845-881.

Sims, S.R., S.A. Berberich, D.L. Nida, L.L. Sagalini, J.N. Leach, C.C. Ebert and R.L. Fuchs. (1996) Analysis of expressed proteins in fiber fractions from insect-protected and glyphosate-tolerant cotton varieties. *Crop Sci.* 36: 1212-1216.

Sims, S.R. and L.R. Holden. (1996) Insect bioassay for determining soil degradation of *Bacillus thuringiensis* subsp. *kurstaki* [CryIA(b)] protein in corn tissues. *Environ. Entomol.* 25: 659-664.

Sims, S.R. and J.W. Martin. (1997) Effect of *Bacillus thuringiensis* insecticidal proteins CryIA(b), CryIA(c), CryIIA, CryIIIA on *Folsomia candida* and *Xenylla grisea* (Insecta: Collembola). *Pedobiologia* 41: 412-416.

Sims, S.R. and J.E. Ream. (1997) Soil inactivation of the *Bacillus thuringiensis* subsp. *kurstaki* CryIIA insecticidal protein within transgenic cotton tissue: Laboratory microcosm and field studies. *J. Agric. Food Chem.* 45: 1502-1505.

Spiegelhalter, F., F.-R. Lauter and J.M. Russell. (2001) Detection of genetically modified food products in a commercial laboratory. *J. Food Sci.* 66: 634-640.

Stone, T.B. and S.R. Sims. (1993) Geographic susceptibility of *Heliothis virescens* and *Helicoverpa zea* (Lepidoptera: Noctuidae) to *Bacillus thuringiensis*. *J. Econ. Entomol.* 86: 989-994.

Sumerford, D.V., D.D. Hardee, L.C. Adams and W.L. Solomon. (1999) Status of monitoring for tolerance to cry1Ac in populations of *Helicoverpa zea* and *Heliothis virescens*: Three-year summary. Proc. Beltwide Cotton Conferences 1999. National Cotton Council, Memphis, TN, pp. 936-939.

Tabashnik, B.E., A.L. Patin, T.J. Dennehy, Y.B. Liu, E. Miller and R.T. Staten. (1999) Dispersal of pink bollworm (Lepidoptera: Gelechiidae) males in transgenic cotton that produces a *Bacillus thuringiensis* toxin. *J. Econ. Entomol.* 92: 772-780.

Tabashnik, B.E., A.L. Patin, T.J. Dennehy, Y.-B. Liu, Y. Carriere, M.A. Sims and L. Antilla. (2000) Frequency of resistance to *Bacillus thuringiensis* in field populations of pink bollworm. *Proc. Natl. Acad. Sci. USA* 97: 12980-12984.

Tapp, H., L. Calamai and G. Stotzky. (1994) Absorption and binding of the insecticidal proteins from *Bacillus thuringiensis* subsp. *kurstaki* and subsp. *tenebrionis* on clay minerals. *Soil Biol. Biochem.* 26: 663-679.

Thayer, A.M. (1999) Ag biotech food: risky or risk free? *C&EN*, November 1, 1999, pp.11-20.

Traxler, G., S. Godoy-Avila, J. Falck-Zepeda and J.J. Espinosa-Arellano. (2001) Transgenic cotton in Mexico: Economic and environmental impacts. <*http://www. biotech-info.net/Bt_cotton_Mexico.html*>.

Umbeck, P.F., K.A. Barton, E.V. Nordheim, J.C. McCarty, W.L. Parrott and J.N. Jenkins. (1991) Degree of pollen dispersal by insects from a field test of genetically engineered cotton. *J. Econ. Entomology* 84: 1943-1991.

Vassiere, B.E. and V.S. Bradleigh. (1994) Pollen morphology and its effect on pollen collection by honeybees, *Apis mellifera* L. (Hymenoptera: Apidae), with special reference to upland cotton, *Gossypium hirsutum* L. (Malvaceae). *Grana* 33: 128-138.

Walker, D.R., H.R. Boerma, J.N. All and W.A. Parrott. (2002) Transgenic technology for insect resistance: Current achievements and future prospects. In: K. Rajasekaran, T.J. Jacks and J.W. Finley (eds.), Crop Biotechnology. ACS Symposium Series No. 829, American Chemical Society, Washington, DC.

Ward, R.N. and K. Ward. (2001) Impact of honeybee pollination activities on Bt cotton production in northern Alabama. Annual Meeting of the Entomological Society of America, Southeastern Branch, Augusta, GA, March 3-7, 2001.

Wilkins, T.A., K. Rajasekaran and D.M. Anderson. (2000) Cotton biotechnology. *Crit. Rev. Plant Sci.* 19: 511-550.

Wraight, C.L., A.R. Zangerl, M.J. Carroll and M.R. Berenbaum. (2000) Absence of toxicity of *Bacillus thuringiensis* pollen to black swallowtails under field conditions. *Proc. Natl. Acad. Sci. USA* 97: 7700-7703.

Xia, J.Y., J.J. Cui, L.H. Ma, S.X. Dong and X.F. Cui. (1999) The role of transgenic Bt cotton in integrated insect pest management. *Acta Gossypii Sinica* 11:57-64.

Yu, L., R.E. Berry and B.A. Croft. (1997) Effects of *Bacillus thuringiensis* toxins in transgenic cotton and potato on *Folsomia candida* (Collembola: Isotomidae) and *Oppia nitens* (Acari: Orbatidae). *J. Econ. Entomol.* 90: 113-118.

Zangerl, A.R., D. McKenna, C.L. Wright, M. Carroll, P. Ficarello, R. Warner and M.R. Berenbaum. (2001) Effects of exposure to event 176 *Bacillus thuringiensis* corn pollen on monarch and black swallowtail caterpillars under field conditions. *Proc. Natl. Acad. Sci. USA* 98: 11908-11912.

Current Resistance Management Requirements for Bt Cotton in the United States

Sharlene R. Matten
Alan H. Reynolds

SUMMARY. The United States Environmental Protection Agency has required an unprecedented insect resistance management (IRM) program for Bt cotton. The specific IRM strategies and requirements for Bt cotton (Bollgard™ cotton) in the United States have been developed by a coalition of stakeholders including EPA, USDA, academic scientists, public interest groups, Monsanto Company, and grower organizations.

Sharlene R. Matten and Alan H. Reynolds are affiliated with the United States Environmental Protection Agency, Office of Pesticide Programs, Biopesticides and Pollution Prevention Division (7511C), 1200 Pennsylvania Avenue, NW, Washington, DC 20460 (E-mail: srmatten@epa.gov).

The authors wish to acknowledge the valuable editorial input of the following individuals: D. D. Hardee, USDA/Agricultural Research Service/Southern Insect Management Research Unit; Nicholas Storer, Dow Agrosciences; and Janet Andersen, Director, USEPA/Office of Pesticide Programs/Biopesticides and Pollution Prevention Division. In addition, the authors wish to acknowledge the contributions of Robyn Rose and other members of the USEPA/Office of Pesticide Programs/Biopesticides and Pollution Prevention Division who helped with materials contributing to this manuscript.

The views expressed in this article are those of the authors and do not necessarily represent those of the United States Environmental Protection Agency (EPA). The use of trade, firm, or corporation names in this article is for the information and convenience of the reader. Such use does not constitute an official endorsement or approval by EPA of any product or service to the exclusion of others that may be suitable.

[Haworth co-indexing entry note]: "Current Resistance Management Requirements for Bt Cotton in the United States." Matten, Sharlene R., and Alan H. Reynolds. Co-published simultaneously in *Journal of New Seeds* (Food Products Press, an imprint of The Haworth Press, Inc.) Vol. 5, No. 2/3, 2003, pp. 137-178; and: *Bacillus thuringiensis: A Cornerstone of Modern Agriculture* (ed: Matthew Metz) Food Products Press, an imprint of The Haworth Press, Inc., 2003, pp. 137-178. Single or multiple copies of this article are available for a fee from The Haworth Document Delivery Service [1-800-HAWORTH, 9:00 a.m. - 5:00 p.m. (EST). E-mail address: docdelivery@haworthpress.com].

http://www.haworthpress.com/store/product.asp?sku=J153
10.1300/J153v05n02_04

There are seven basic IRM requirements: structured refuge, resistance monitoring, remedial action plan, compliance assurance, grower education, grower agreements, and annual reports. The mandatory refuge requirements for Bollgard cotton are: (1) 5% external, unsprayed refuge option, (2) 20% external, sprayed refuge option, (3) 5% embedded refuge option, or (4) a community refuge option that may utilize the 5% external, unsprayed refuge option and/or the 20% external, sprayed refuge option. Each refuge option also has specific deployment requirements. The current registration for Cry1Ac plant-incorporated protectant as expressed in cotton (known as Bollgard cotton) expires September 30, 2006, except for the 5% external, unsprayed structured refuge option, which will expire September 30, 2004. Additional research data will be integral to strengthening the existing IRM plans for Bollgard cotton and promoting greater sustainability. *[Article copies available for a fee from The Haworth Document Delivery Service: 1-800-HAWORTH. E-mail address: <docdelivery@haworthpress.com> Website: <http://www.HaworthPress.com>]*

KEYWORDS. Insect resistance management (IRM), refuge, integrated pest management (IPM), susceptible, plant-incorporated protectant

INTRODUCTION

The United States Environmental Protection Agency (EPA) has required an unprecedented insect resistance management (IRM) program as part of the Bollgard™ cotton (*Gossypium hirsutum*) registration since it was first registered in 1995. This technology consists of Cry1Ac from *Bacillus thuringiensis* (Bt) as a plant-incorporated protectant (PIP) expressed in transgenic cotton. Under the Federal Insecticide, Fungicide, and Rodenticide Act (FIFRA), the United States Environmental Protection Agency (EPA) ensures there will be no unreasonable, adverse effects from the use of a pesticide when economic factors are fully considered. The law allows EPA to require pesticide registrants to develop and implement insect resistance management (IRM) plans. EPA considers IRM for Bt crops such as cotton to be very important, and the maintenance of insect susceptibility to Bt to be in the "public good."

IRM describes practices aimed at reducing the potential for insect pests to become resistant to a pesticide. The Agency has determined that development of resistant insects would constitute an adverse environmental effect. Thus, EPA is working to prevent substitution of more toxic insecticides in the place of Bt if Bt became unable to control insect pests. Academic and government scientists, public interest groups, and organic and other farmers have expressed concern that the widespread planting of these genetically modified plants will hasten insect resistance to Bt. Sound IRM will prolong the life of Bt insecticides, and adherence to IRM plans benefits growers, producers, researchers, and consumers. EPA's strategy to address insect resistance to Bt is twofold: (1) mitigate any significant potential for pest resistance by instituting IRM plans; and (2) better understand how pest resistance happens and how it can best be stopped.

The Cry1Ac plant-incorporated protectant as expressed in cotton (Bollgard cotton) is toxic to certain lepidopteran insect pests, in particular, *Helicoverpa zea* (Boddie) (cotton bollworm, CBW), *Heliothis virescens* (Fabricius) (tobacco budworm, TBW), and *Gossypiella pectinophora* (Saunders) (pink bollworm, PBW). Beginning in 1995, the Agency determined that the unrestricted use of Bt cotton would likely lead to the emergence of resistance in one or more of the target insect pests unless measures are used to delay or halt development of resistant insects. Because some cotton pests also attack other crops, not only would the emergence of resistance affect the benefits of Cry1Ac cotton, such insect resistance could also affect the efficacy of Bt corn products and some microbial formulations of Bt. The loss of Bt as an effective pest management tool–in cotton or other crops–could potentially have serious adverse consequences for the environment to the extent that growers would shift to the use of more toxic pesticides, and a valuable tool for organic farmers could be impacted. The emergence of resistance in cotton pests could also have significant economic consequences for cotton growers. Therefore, EPA required Monsanto (the registrant) to implement an IRM program to mitigate the possibility that pest resistance will occur. This IRM program for Bollgard cotton is unprecedented, detailed, and proactive as compared to any other registered synthetic pesticide used on cotton or any other crop.

The development of scientifically sound and sustainable IRM strategies for Bt cotton has required a number of stakeholders (growers, seed suppliers, scientists, and regulators) to be engaged in the process. The issue of IRM for Bt crops, in general, has generated more data, meetings, and public comments in favor of proactive, long-range protection of a pesticidal active ingredient than for all other pesticides combined. Because of the importance of protecting Bt, EPA has taken several very important steps to work with stakeholders. EPA has held multiple fora that have focused on insect resistance management for Bt crops: three

Federal Insecticide Fungicide and Rodenticide Act Scientific Advisory Panel (SAP) Meetings (1995, 1998, 2000), six public workshops (1999, 2001), two public hearings (1997), two Office of Pesticide Program Dialogue Committee Meetings (1996, 1999), and one technical briefing (2001). In addition, EPA has worked with USDA, Monsanto (the sole registrant at present for Bt cotton hybrids under the trademark Bollgard™) and cotton producer organizations: the National Cotton Council, the Arizona Cotton Research and Protection Council, and the Delta Council. Agency staff have presented papers at the Entomological Society of America annual meetings and Beltwide Cotton Conferences on Bt cotton IRM (and other Bt crops). In addition, Agency staff have visited with many cotton producers on their farms to discuss Bt cotton IRM issues. These meetings have provided opportunities for sharing information, establishing research priorities, and building trust among participants. The overall goal of all of these stakeholder fora has been to identify the best science-based, yet practical, IRM strategies. The USDA has also published a document that explains how Bt cotton is developed, how it controls insect pests, how it can most effectively be used in insect pest management programs including the use of insect resistance management strategies to preserve the technology (Hardee et al. 2000). Producer participation in IRM meetings has been particularly valuable because their interest in Bt cotton is high and they often bring common-sense approaches to the forefront.

   IRM for Bt crops has focused on the use of the high-dose/structured refuge strategy to mitigate insect resistance Bt. Scientists believe that a high dose and the planting of a refuge (a portion of the total acreage using non-Bt plants) delays the development of insect resistance to Bt crops by maintaining insect susceptibility. This IRM approach assumes that the pest biology and ecology are well known and that the Bt crops are used in a way that complements an overall integrated pest management (IPM) plan. In addition to a high dose and structured refuge, IRM plans include additional field research on pest biology, refuge size and deployment, resistance monitoring for the development of resistance, grower education and compliance, a remedial action plan in case resistance is identified, annual reporting and communication. IRM plans will change as more scientific data become available.

   This chapter considers IRM for Bt cotton (Bollgard cotton) with an emphasis on (1) IRM program requirements, (2) dose expression and structured refuge, (3) resistance management modeling, (4) structured refuge requirements, (5) resistance monitoring requirements, (6) remedial action plans, (7) grower education and compliance, and (8) research data requirements.

Following the registration of Bt cotton, acreage planted in the US (Table 1) has steadily increased between 1995 and 2001.

IRM PROGRAM REQUIREMENTS

The current registration for Cry1Ac plant-incorporated protectant (PIP) as expressed in Bt cotton (Bollgard) expires September 30, 2006 except for the 5% external, unsprayed structured refuge option which will expire September 30, 2004.

The required IRM program for Bt cotton has the following general elements:

1. Requirements relating to creation of a non-Bt cotton refuge in conjunction with the planting of any acreage of Bt cotton;
2. Requirements for the registrant to prepare and require Bt cotton users to sign "grower agreements" which impose binding contractual obligations on the grower to comply with the refuge requirements;
3. Requirements for the registrant to develop, implement, and report to EPA on programs to educate growers about IRM requirements;
4. Requirements for the registrant to develop, implement, and report to EPA on programs to evaluate and promote growers' compliance with IRM requirements;
5. Requirements for the registrant to develop, implement, and report to EPA on programs to evaluate whether there are statistically significant and biologically relevant changes in susceptibility to Cry1Ac protein in the target insects;
6. Requirements for the registrant to develop, and if triggered, to implement a "remedial action plan" which would contain measures the registrant would take in the event that any insect resistance was detected as well as to report on activity under the plan to EPA;
7. Requirements to submit annual reports on resistance monitoring, compliance, research, and sales by January 31st each year.

DOSE EXPRESSION AND STRUCTURED REFUGE

Several IRM strategies to minimize selection pressure for resistant insects on Bt crops have been proposed. The strategy that has received the most attention involves two components: high dose and structured refuge (Roush 1997a,b, Gould 1998a,b). The high-dose/refuge strategy assumes that resistance to Bt plants is recessive and is conferred by a single locus with two alleles resulting in three insect genotypes: susceptible homozygotes (SS), heterozygotes (RS), and resistant homozygotes (RR) (Roush 1994, Gould 1998b, USEPA 1998, Bourguet et al. 2000). It also assumes that there will be a low initial frequency of resistance alleles and that there will be random mating between resistant and susceptible adults. Under ideal circumstances, only rare RR individuals will survive a high dose produced by the Bt crop, and both SS and RS individuals will be susceptible to Bt. A structured refuge is a non-Bt portion of a field or set of fields that provides for the maintenance of susceptible (SS) insects. Susceptible insects (SS) from the refuge will mate with rare resistant RR insects surviving the Bt crop to produce susceptible

TABLE 1. Total Cotton Acreage Planted to Bt Cotton (Bollgard™) Hybrids in States

State	Bollgard™ Acreage					
	1996	1997	1998	1999	2000	2001
Alabama	348,810	251,784	306,535	398,683	314,500	255,777
Arizona	53,290	175,537	207,713	197,911	210,245	244,409
Arkansas	166,881	113,490	111,818	173,652	294,364	612,266
California	618	9,868	29,129	91,705	54,584	76,161
Florida	52,836	55,030	53,377	45,249	48,974	51,976
Georgia	375,744	533,340	508,842	693,288	580,908	568,087
Kansas	-	-	-	-	1,056	304
Kentucky	-	-	-	-	-	980
Louisiana	157,411	202,080	244,616	382,839	450,076	713,605
Mississippi	443,986	410,333	506,149	746,163	800,775	1,247,740
Missouri	498	592	519	6,254	21,415	88,448
New Mexico	393	2,693	20,869	12,263	12,242	16,393
North Carolina	20,519	21,027	77,490	274,312	424,880	543,888
Oklahoma	11,772	7,103	11,459	69,545	90,925	131,759
South Carolina	53,864	91,891	71,894	176,149	128,684	131,362
Tennessee	10,833	17,431	57,649	390,245	380,453	476,061
Texas	98,819	186,654	276,520	458,694	570,410	608,118
Virginia	86	37	1,876	6,300	24,857	36,490
US Total	1,796,390	2,078,890	2,486,493	3,585,437	4,409,348	5,803,824

(RS) heterozygotes that will be killed by the Bt crop. This strategy should dilute resistant (R) alleles from the insect populations and delay the evolution of resistance.

The SAP recognized that resistance management programs should be based on the use of both a high dose of Bt and structured refuges designed to provide sufficient numbers of susceptible adult insects (SAP 1998). The 1998 and 2000 SAP subpanels noted that IRM strategies should also be sustainable and strongly consider grower acceptance and logistical feasibility (SAP 1998, 2001).

Although the high dose/refuge strategy is the preferred strategy for IRM, effective IRM is still possible even if the transformed plant does not express the Bt protein at a high dose to all economically-important target pests. Although the high-dose/refuge strategy is the preferred strategy for IRM in Bt crops, effective IRM may still be possible under non-high dose conditions, for example, by increasing refuge size, or limiting total acres. The lack of a high dose could allow partially resistant (i.e., RS insects) to survive, thus increasing the frequency of resistance genes in an insect population. For this reason, numerous IRM researchers and expert groups have concurred that non-high dose Bt expression presents a substantial resistance risk relative to high dose expression (Roush 1994, Gould 1998a, 1998b, Onstad and Gould 1998, SAP 1998, ILSI 1998, Mellon and Rissler 1998, SAP 2001).

High Dose

The SAP (SAP 1998) noted that a Bt PIP could be considered to provide a high dose if verified by at least two of the following five approaches: (1) serial dilution bioassay with artificial diet containing lyophilized tissues of Bt plants plus tissues from non-Bt plants as controls; (2) feeding assays using plant lines with expression levels approximately 25-fold lower than the commercial cultivar determined by quantitative enzyme linked immunosorbent assay (ELISA) or some more reliable technique; (3) survey large numbers of commercial plants in the field to make sure that the cultivar is at the lethal dose $(LD)_{99.9}$ or higher to ensure that 95% of heterozygotes would be killed (Andow and Hutchison 1998); (4) similar to approach 3 but would use controlled infestation with a laboratory strain of the pest that had an LD_{50} value similar to field strains; and (5) determine whether a later instar of the targeted pest could be found with an LD_{50} that was approximately 25-fold higher than that of the neonate larvae. If so, the later stage could be tested on the Bt crop plants to determine whether 95% of the larvae were killed. EPA has adopted the 25X definition of "high dose," but agrees with the 2000 SAP Subpanel that this definition is "imprecise, provisional, and may require modification as more knowledge becomes available about the inheritance of resistance" (SAP 2001).

All Bt cotton cultivars in the U.S. probably produced a high dose for TBW and PBW, while none of the cultivars produce a high dose (instead a moderate dose) for CBW (SAP 1998, 2001, USEPA 2001). With CBW, there is greater than 20% survival from exposure to the level of Cry1Ac present in the current product (EPA 1998, Mahaffey et al. 1994). Gould (1998b) noted that if a moderate dose is to be sustainable then the refuge size should be significantly larger, but some of the spatial requirements may be less strict then they are for the high dose. The 2000 SAP subpanel agreed with Gould's analysis for CBW that refuge size is more important than spatial requirements for the refuge. Gould and Tabashnik (1998) cited the need for very large (30 to 50%) refuges to manage CBW resistance based on modeling results. With moderate dose Bt crops, "larval movement is not expected to reduce recessiveness significantly." However, Gould (1998b) also stated that wild hosts and other crops could serve as part of a larger refuge for CBW, but that data on the contribution of these hosts to overall pest population size in different geographic areas is lacking.

An effective insect resistance management strategy for Bt cotton, therefore, must consider the differential effect of having a high dose for TBW and PBW, but only a moderate dose for CBW. The 2000 SAP Subpanel stated that for CBW, the amount of refuge in a region (i.e., percentage of non-Bt cotton in an area) is more important than refuge deployment on individual farms (i.e., refuge proximity and structure) because of the long-distance movement of CBW and the lack of a high dose (SAP, 2001).

Refuge

The size, placement, and management of refuge plantings are critical to the success of a high-dose/structured refuge strategy in mitigating insect resistance to Bt. Structured refuges include all suitable non-Bt host plants for target pests (SAP 1998). A 500:1 ratio of susceptible-to-resistant insects has been suggested as a suitable goal, assuming a resistance allele frequency of 5×10^{-2} (SAP 1998, 2001). Planting date, size, and placement of structured refuge should be based on current pest biology and should maximize the overlap of susceptible insects with possible resistant insects.

CBW larvae, particularly later instars, are capable of plant-to-plant movement. For this reason, seed mixes do not appear to be viable refuge option for CBW (SAP 1998). However, PBW has very limited larval mobility and seed mixes may be a viable option (Watson 1995, USEPA 1998, 2001).

TBW and CBW are polyphagous insects, feeding on a number of grain and vegetable crops in addition to weeds and other wild hosts (described in Caprio and Benedict 1996, Stadelbacher et al. 1986, USEPA 1998). The Agency is aware that pests such as CBW have alternate hosts; CBW when feeding on corn is known as corn ear worm (CEW). There is little if any empirical evidence to support the inclusion of these as effective refuge. This conclusion is supported by both the 1998 and 2000 SAP subpanels (SAP 1998, 2001). The February 1998 SAP (SAP 1998) was asked by the Agency to consider the utility of alternate hosts for CBW as refuge. They concluded that, "until it is shown that non-cotton hosts produce enough susceptible moths to significantly delay the evolution of resistance in CBW/CEW populations exposed to moderate Bt doses, non-Bt cotton acreage must be considered the primary source of susceptible CBW/CEW moths."

Similarly, the October, 2000 SAP subpanel concluded that there was not enough empirical evidence to support alternate hosts as efficient refuge. They noted that CBW resistance models do not consider soybean as a refuge because most information indicates that it isn't a reliable host each season. They state: "A best case scenario model would include soybean as a refuge. If there were better empirical data on soybeans, a more realistic model could be developed that accounted for the true year to year variation in the utility of soybean as a refuge" (SAP 2001). Gould and Tabashnik (1998) also indicated in their analysis of Bt cotton IRM plans "with soybean and corn, more research is needed to determine how many useful adult insects are produced per acre at different locations during several years." Therefore, based on evidence provided to the Agency, until such time as there is sufficient empirical data that demonstrate that alternate hosts are producing insects in sufficient quantity, temporal synchrony, fitness, and proximity to the resistant insects that would be emerging from Cry1Ac Bt cotton fields, then only non-Bt cotton can be used as an efficient refuge. This uncertainty prompted EPA to require the registrant to further study alternate hosts as refuge.

Recent research supports the possibility that alternate hosts such as corn may provide functional refugia for Bt-susceptible insects (Gould et al., 2002), but much more data will be necessary before alternate hosts can be incorporated into IRM planning.

Non-Bt cotton hybrids used as refuges should be selected for growth, maturity, fertility, irrigation, weed management, planting date, and yield traits similar to the Bt cotton hybrid. Hybrids that are not agronomically similar may result in different developmental times in cotton pests that could lead to assortative (non-random) mating between insects from the refuge and Bt cotton fields.

Laboratory studies have shown that temporal mating can potentially occur among PBW populations from non-Bt and Bt cotton, since development is delayed for resistant larvae feeding on Bt cotton (Liu et al. 1999). Peck et al. (1999) have shown in model simulations for TBW that interactions between developmental asynchrony and season length increase uncertainty because they either hasten or slow the evolution of resistance. However, it is important to note that there is considerable overlap in generations of this insect occurring in the field, especially late in the season (Shelton and Roush 1999). Asynchronous development may have either negative or positive effects on the effectiveness of the high dose/refuge strategy in the field; therefore, further study is warranted.

Proximity and Refuge Deployment

Refuge proximity and deployment are critical variables for resistance management. Refuges must be located so that the potential for random mating between susceptible moths from the refuge and any resistant survivors from the Bt cotton field is maximized. Therefore, insect flight and ovipositional behavior are critical variables to consider for refuge proximity. Refuges planted as external blocks should be adjacent or in relatively close proximity to the Bt cotton fields.

There are trade-offs when considering refuge placement that involve movement of insect larvae and adults. In general, proximity of Bt and refuge plants benefits random mating of adults, but it increases the chances that larvae may move between Bt plants and refuge plants. Distant and temporal plantings of Bt and refuge plants eliminate problems with larval movement, but potentially compromise random mating of susceptible and resistant adults. Caprio (2001) has modeled the interaction of population dynamics and movement among Bt and non-Bt refuge fields on the evolution of insecticide resistance. His models suggest that resistance management programs using untreated non-Bt refuges should not overly focus on random mating at the cost of making the habitat too fine-grained (e.g., in-field strips of non-Bt cotton that are very narrow in width). Refuge configurations that have been considered include external blocks, in-field strips/blocks, seed mixes, temporal refuge, and non-cotton hosts.

TBW and CBW

Published data indicate that both CBW and TBW are highly mobile insects, capable of significant long distance movement, with CBW being more mobile than TBW. Mark/recapture studies have shown that CBW moths can disperse distances ranging from 0.5 km to 160 km, and in rare instances up to 750 km (Caprio and Benedict 1996).

The general pattern of migration is a northward movement, following prevailing wind patterns, with moths originating in southern overwintering sites moving to corn-growing regions in the northern US and Canada. Observations based on carbon isotope studies indicate that CBW may also move southward from corn-growing regions back to cotton regions in the South: 48-72% of the moths collected in Louisiana and Texas cotton fields in late September and early October developed on C4 plants (e.g., corn, sorghum, other grasses) (USEPA 2001, SAP 2001, Gould's remarks in USEPA/USDA 1999). If further investigation confirms this is true, the result may be additional CBW exposure to Bt crops. In addition, the assumptions about CBW overwintering may need to be revisited–moths that were thought to be incapable of winter survival (and thus not a resistance threat) may indeed be moving south to suitable overwintering sites. The CBW moths in question may actually have come from sources other than northern corn, such as grain sorghum or C4 weeds. The general consensus of the 2000 SAP was that southward CBW migration was not proven, but that there was considerable circumstantial evidence for it. They concluded that potential southward movement should be considered in resistance management. The SAP recommended further scientific investigation of CBW reverse migration which prompted EPA to require the registrant to further study this area.

Movement at a localized level may be more important for the design of a refuge because of the need for random mating and oviposition. The 1998 SAP subpanel noted that research has shown that substantial local population substructure can develop during the summer as a result of restricted localized movement of TBW and, therefore, deployment of a structured refuge is important (SAP 1998). Because of this, Gould and Tabashnik (1998) recommended that the distance between Bt cotton fields and the non-Bt cotton refuge should be a maximum of 1.67 kilometers.

Based on ovipositional patterns for CBW, Caprio (2000a) has indicated that untreated, embedded refuges should be at least 100 m wide to minimize the risk of rapid resistance evolution associated with source-sink dynamics. The refuge must be wide enough so that all females do not lay all of their eggs in the Bt portion of a field, and close enough to the Bt portion of the field so that mating and oviposition of adults can be random. A spatial restriction for the refuge of less than one km may be more appropriate for CBW based on his movement studies.

PBW

Studies of PBW in non-Bt cotton show that some adults disperse long distances, but most do not (see discussion in Tabashnik et al. 1999, Carrière et al. 2001). Tabashnik et al. (1999) measured male dispersal at a single site of 259 ha (1 mile^2) containing 69% Bt cotton and 31% non-Bt cotton. The distribution of wild males caught in pheromone traps suggested that many moved at least 400 m from non-Bt cotton to Bt cotton; yet, the movement was not sufficient to achieve a random distribution of males between non-Bt and Bt cotton. Using sterile males, 66-94% dispersed 400 m or less from the release sites.

Carrière et al. (2001) estimated dispersal distances of PBW by tracking movement of males and females from isolated non-Bt cotton refuges (source) into surrounding Bt cotton (sink). They noted that because Bt cotton acts as a deadly sink, most moths flying in Bt cotton at the end of the growing season (September-November) must originate from refuges. Their results showed that dispersal of females from non-Bt cotton to Bt cotton was dramatically reduced at only 0.83 km from the border of the refuge.

Together the results from both male and female dispersal experiments (Tabashnik et al. 1999, Carrière et al. 2001) and previously published data on PBW dispersal suggest that refuges for PBW should be close to Bt cotton to promote random mating of resistant and susceptible individuals. Both the Tabashnik et al. (1999) and Carrière et al. (2001) groups recommended that the distance between Bt cotton fields and refuges should be no more than 1.67 km to favor mating between the RR (resistant individuals) from the Bt cotton fields and the SS (susceptible) from the refuges. A more precise evaluation of the effect of size and distance of refuges on the number of moths dispersing to Bt cotton fields is being conducted by Carrière and his research group in Arizona.

Because PBW larval movement is limited between plants, Gould and Tabashnik (1998) recommended in-field refuges as the best approach to PBW resistance management because "they reduce or eliminate the isolation by distance that could reduce hybrid matings between susceptible adults from refuges and resistant adults from Bt cotton." Gould (1998b) states that "a within-field refuge (e.g., seed mixture or row by row mixing) would be best because of limited larval and adult movement." Therefore, PBW adult dispersal information support placement of the refuge as close to the Bt cotton fields as possible to maximize random mating. Because of the limited larval movement, EPA has allowed narrow in-field strips (at least one row) of non-Bt cotton to be planted for every six to ten rows of Bt cotton in the same field (USEPA 2001).

RESISTANCE MANAGEMENT MODELING

EPA has used predictive models to compare IRM strategies for Bt cotton. In the absence of field resistance and without a complete understanding of the biology of the pest, all models depend on the use of some assumptions. The assumptions differ among models and can greatly affect the output of the model. Predictive models used for resistance management are very sensitive to assumptions about the genetics of field-selected resistance (e.g., gene frequency and functional dominance) about which little is known for most cases.

Deterministic models exclude random events. The same input will always produce the same output. Under discrete, non-random conditions, deterministic models may examine resistance evolution within a single field or several thousand fields. In contrast, stochastic, spatially explicit models incorporate both random events and a spatial dimension. In stochastic models, a given input may produce many different outputs because some of the parameters in the model are random variables. These models can be used to look at resistance evolution in multiple fields or patches used by multiple growers (a regional approach) and consider the impact of population dynamics and population structure. Five Bt cotton insect resistance management models used by EPA are discussed below.

Gould's Deterministic Model
for TBW and CBW Resistance Management

Gould modeled the performance of several refuge scenarios (Gould 2000, Table 2). This deterministic model assumes diploid genetics, ran-

TABLE 2. Gould's Deterministic Model for TBW and CBW Resistance Management

Fitness of Bt plants*	Years to Resistance Allele Frequency Reaching 0.50 for Varied Refuge Sizes (Unsprayed)			
	4% refuge	5% refuge	10% refuge	20% refuge
Case 1: Extremely high efficacy against susceptible insects RR = 1.0; RS = 0.01; SS = 0.0001	5.3	6.3	11.0	22.7
Case 2: Very high efficacy against susceptible insects RR = 1.0; RS = 0.01; SS = 0.001	5.7	6.7	11.7	24
Case 3 [Case for TBW]: Extremely high efficacy against susceptible insects RR = 1.0; RS = 0.001; SS = 0.0001	12	14.7	29	62.3
Case 4 [Case for TBW]: Very high efficacy against susceptible insects RR = 1.0; RS = 0.002; SS = 0.001	12	14.7	28.3	61
Case 5: Moderate/high efficacy against susceptible insects RR = 1.0; RS = 0.02; SS = 0.01	6	7	12	23.3
Case 6 [Case appropriate for CBW]: Moderate efficacy against susceptible insects RR = 1.0; RS = 0.2; SS = 0.1	4	4.3	5.3	7.7

* RR = (homozygous resistant); RS = (heterozygous); SS = (homozygous susceptible)

dom mating, three generations per year, and initial resistance allele frequency of 0.001, while excluding density dependence and spatiality. Gould varied the degree of mortality of susceptible larvae to account for crops with differing compatibility with the high dose concept. He also varied the degree of recessiveness of the resistance alleles. All scenarios were for external unsprayed refuge options. This model is conservative and represents a worst-case projection. Results of this model point to the need for more detailed models that include population dynamics and population structure.

Caprio's Spatial, Stochastic Model for CBW Resistance Management

Caprio modeled the effect of different refuge scenarios on CBW resistance (Caprio 2000b, 2000c, 2001). Caprio's model assumed CBW being exposed exclusively to cotton through four generations/year. Most areas will have a substantial refuge in corn during the first two generations, so this model might represent a worst case (depending on whether or not Bt corn is growing in the area). This is not an unlikely one, as there is not a great deal of corn grown in much of the Cotton Belt. In the model, he assumed 5% survivorship of susceptible larvae, 0.002 initial gene frequency, and that resistance is a partially recessive trait (h = 0.1). Overwintering survival was estimated to be 25%. Dispersal associated with overwintering and the first spring generation (from non-crop hosts to cotton) was assumed to be 90%. This estimate was probably low, but was used to overcome scale limitations associated with complex simulations. The daily dispersal rate for the first two generations on crop hosts was assumed to be 80%/day. It is assumed that cotton is not a very good host during this time, and CBW moves from field to field. Refuges are assumed to be in the same location each year. However, Caprio noted that this would not be a problem given the high dispersal during overwintering and the first two generations. Wild hosts are not simulated. For the last two generations, dispersal is set at 25%/day (i.e., 25% of adults leave a patch per day–a field may consist of many 10-acre patches). Caprio calculated that about 46% of the eggs from females emerging in the refuges are laid in the refuge. With dispersal set to 50% per day, 21% of eggs from females emerging in the refuges are laid in the refuge. This is about what Caprio estimated for refuges that are approximately 100 m wide (67% dispersal parameter). Larval movement is ignored in this model. The number given by the model is years until 50% of the fields have resistance allele frequencies above 50% (Table 3).

These simulations predict that a seed mix or single rows would not be used to effectively manage CBW resistance. To delay resistance more than 10 years, Caprio's model indicates there would have to be greater than 30% non-Bt cotton in a (untreated) seed mix. His simulations predict that embedded (untreated) options give the greatest benefit for resistance management with the least amount of non-Bt cotton planted, assuming the refuge is wide enough to create sufficient isolation between the refuges and Bt fields. This isolation ensures that females from refuges lay enough eggs in refuges to maintain a large susceptible population in those areas. At the same time, the refuge should be close enough to the Bt portion of the field so that the increased isolation does not lead to an increase in non-random mating that could overcome the effectiveness of the non-random oviposition. Overall, these simulations suggest there must be a balance between isolation limiting source-sink effects (and delaying resistance evolution) and isolation increasing non-random mating (and hastening resistance evolution) (also see Caprio

TABLE 3. Caprio's Model for *H. zea* (CBW) Resistance Management

Refuge Option	Years to Resistance
Untreated (more like a seed mix or single row)	
4%	3.46 years (+ 2 extinctions)
16%	5.3 years (+ 2 extinctions)
32%	9.5 years
Sprayed external refuges (economic threshold at 4% with 90% efficacy of the larval population)	
0%	2.2
10%	7.25
20%	10.5
30%	14.5
Embedded untreated refuges (50% Dispersal)	
1.25%	8.6
2.5%	10.3
5.0%	19.2
10.0%	24.8
Embedded untreated refuges (67% Dispersal)	
1.25%	7.0
2.5%	8.0
5.0%	12.0
10.0%	22.4

2001). These simulations did not consider the influence of alternate hosts, such as corn, on the development of CBW resistance to Bt.

Livingston's Efficient Refuge Model

The model developed by Livingston et al. (2002) seeks to maximize grower profits while achieving an economically acceptable rate of resistance evolution. This model differs from strictly biological models that predict the number of years until resistance occurs (resistance allele frequency reaches a certain level) under different refuge scenarios.

Livingston et al. (2002) derived sprayed and unsprayed refuges that maximize the present value of average profits per acre received by representative Louisiana (or mid-South) cotton producers. The economic model incorporated genetic models of Bt cotton and pyrethroid resistance evolution in TBW and CBW. Sprayed and unsprayed refuges maximize grower profitability subject to producer behavior and resistance evolution over an eleven-year planning horizon. Refuges were derived from two- through eleven-year horizons, initiated at the beginning of the 2000 growing season. The first season of Bt cotton introduction was 1996. Thus, the total time horizon covers a 15-year period from the time in which Bt cotton was first commercialized. Alternative cultivated and non-cultivated hosts for TBW and CBW were considered as part of the model.

Sprayed refuges were insensitive to the interest rate, pest-free yield, output price, and technology fees. Sprayed refuges increased with the time horizon and varied considerably with genetic parameters that influenced Bt resistance evolution in TBW, but not with parameters that influenced pyrethroid resistance in either TBW or CBW. Unsprayed refuges increased with the time horizon, but less so than sprayed refuges, and were sensitive to several economic and genetic parameters.

The producer received higher profits under sprayed relative to unsprayed refuges, and average sprays were lower under unsprayed refuges. Both species became less resistant to Bt under sprayed relative to unsprayed refuges, and both species became more resistant to pyrethroids under sprayed relative to unsprayed refuges. Producer use of unsprayed refuges was not recommended under the default model.

The most important determinants of sprayed refuges were the time horizon and genetic parameters that characterized Bt resistance evolution in the TBW. In order of decreasing importance, these factors were: degree of dominance, initial resistance allele frequency, susceptible fitness, and fitness cost. If the degree of dominance is closer to 0.05, for example, producers could make five and six percent more per acre under fourteen and nine percent sprayed and unsprayed refuges, respectively, over a five-year horizon (initiated at the beginning of the 2000 growing season) relative to earnings under the default degree of dominance (d = 0.1 for Bt).

Unsprayed refuges generally increased with the time horizon but did not increase beyond 11% between the four- and eight-year horizons because pyrethroid susceptibility in the TBW was a renewable resource under the default model. The more important determinants of unsprayed refuges were, in descending order: harvested yield per unsprayed refuge acre, output price, initial frequency, degree of dominance, the time horizon, susceptible fitness, the technology fee, interest rate, and fitness cost. The representative producer earned higher returns, used less insecticides per acre, and managed Bt resistance more effectively under sprayed relative to unsprayed refuges (although average insecticide treatments were lower for unsprayed refuges).

Producer advantages of using sprayed relative to unsprayed refuges declined with harvested yield per unsprayed refuge acre. Producer use of unsprayed refuges was not recommended for producers who harvest less than 600 pounds per unsprayed refuge acre; however, use of unsprayed refuges was recommended for producers who harvest 600 pounds or more per unsprayed refuge acre.

Results of the Livingston et al. (2002) model indicated that the wide host range of CBW had two impacts on Bt resistance evolution. First, the small proportion of the CBW in cotton during the early season increased the refuge size. Second, a significant proportion of CBW remained in non-cotton hosts even when the majority of the CBW populations were in cotton. This also increased its effective refuge size. CBW inherited Bt resistance as a partially dominant trait, with d = 0.75 based on the estimated resistance allele frequency measured by Burd et al. (2001).

If the model is run for 5 years (spanning the end of the 9th year of commercial Bt cotton use), the sprayed refuge size would be 21% and the unsprayed refuge size would be 11% for both TBW and CBW based on average producer profits, Bt and non-Bt costs, and resistance evolution. After 11 years (spanning the end of the 15th year of commercial Bt cotton use), the model predicted that the sprayed refuge size would be 41% and the unsprayed refuge size would be 14%. A brief summary is presented in Table 4.

Peck's Spatial, Stochastic TBW Model

Peck et al. (1999) developed a stochastic, spatially-explicit, simulation model to examine the factors that may influence the regional development of TBW resistance to Bt (Cry1Ac) produced in transgenic cotton. The model follows the fate of a meta-population of TBW arrayed in a grid representing individual agricultural fields. The region size is 25 to 1,250 fields. The model has a daily time step. The initial gene frequency (q^0) is unknown, but the q^0 set was to 0.03 because lower values caused extinctions of the resistance alleles due to the stochastic nature of the model or took too long for the computer simulation to run. The selection coefficients on Bt were: RR = 1, RS = 0.02, SS = 0.01. The refuge size in seed mixtures was 20% and the field-level refuge size was 20%. There was no larval movement and no mortality of moving larvae. Migration was 0.045 per generation. The initial population size in the null run was 60,000 per generation. The K (spraying

TABLE 4. Predicted Efficient Refuges, Sprayed and Unsprayed and Grower Profits*

Time Horizon ($T^0 = 2000$) [Year 5 after 1st commercialization]	Sprayed Refuge	Annualize Present Value Profit per Acre	Unsprayed Refuge	Annualize Present Value Profit per Acre
2†	5%	$269	4%	$263
3	11	261	8	247
4	17	253	11	234
5	21	246	11	222
6	26	240	11	214
7	29	234	11	207
8	33	228	11	200
9	36	223	13	194
10	39	218	14	189
11	41	214	14	184

*Adapted from Livingston et al. 2002.
†Through the end of 2001.

threshold) was 150,000. The R_0 (reproductive potential) was 9. The pesticide mortality in non-Bt fields was 90% and the daily adult mortality was 15%. Winter survival of pupae was 5% and a year occurs every 120 days. The maximum adult movement distance was three fields each day. Resistance was defined in three ways: (1) when 50% of the transgenic fields have a resistance allele frequency of > 50%; (2) when 50% of the transgenic fields have reached a resistance allele frequency of > 90%; and (3) when 50% of the transgenic fields reach 60% of their spraying threshold (the maximum population size a field can sustain before economic damage occurs).

Using this model, Peck et al. (1999) found that the spatial scale and the temporal pattern of refuges can have a strong effect on the development of TBW resistance to Bt cotton. Specifically, the time to resistance was significantly longer (49 years) in regions where the same fields were used as a refuge from year to year and adult movement among fields is limited. In regions where the refuge fields are changed randomly from year to year, the region develops resistance more quickly (17 years). Peck et al. (1999) concluded that it would only take a minority of growers who do not employ refuges properly to start a regional resistance problem. These authors found that 20% (sprayed) refuges did delay resistance. Peck et al. (1999) noted that a delay in larval development on Bt plants can alter the rate of resistance development to increase or decrease the rate of resistance development. They commented that designing controls to limit the overwintering potential of the last generation may be effective in slowing resistance. Exploring the interaction among parameters is very difficult with this complex model, but this type of model is useful to examine a number of challenges to managing resistance in Bt cotton (e.g., how the refuge is managed year to year) and the scale (regional level) of management of resistance. Neither the spatial scale nor temporal pattern of placement of refuges has been investigated in the field. Further study is recommended.

Storer's Spatial, Stochastic H. zea Model

The spatial, stochastic computer model of Storer et al. (1999) was developed to simulate the evolution of resistance in CBW to Bt cotton in an agroecosystem that includes both Bt corn and Bt cotton, such as eastern North Carolina. This model is adapted from that developed by Peck et al. (1999) for simulating resistance evolution in TBW. The model has multiple fields of 10 acres each for a total of 5,260 or 9,000 acres. The proportion of the region planted to corn was 55%, with the remainder planted to cotton. As a default, 75% of corn fields and 25% of cotton fields were planted to Bt varieties (other ratios were also tested). The initial frequency of resistance alleles was assumed to be 10^{-4}. The functional dominance of resistance alleles was assumed to be 0.5 (based on Gould et al. 1995). The susceptible survival on Bt cotton was 25% and the proportion of the region as corn was 55%. The percentage adoption of Bt corn and Bt cotton was 100%.

Using this model, Storer et al. (1999) found that selection for resistance is more intense in Bt cotton fields than in Bt corn fields. For example, when 75% of cotton is Bt cotton and 25% of corn is Bt corn, the R-allele frequency increased more rapidly than if 25% of cotton is Bt cotton and 75% of corn is Bt corn. Storer et al. (1999) concluded that the greater importance of Bt cotton with regard to resistance development was due to spraying of non-Bt cotton fields when they reached economic threshold levels, thereby reducing the effective refuge size. The spatial distribution of transgenic and non-transgenic plantings can affect both the region-wide evolution of resistance and, especially when the on-farm refuge size is small, the resistance levels in sub-populations. Storer et al. (1999) concluded that farm-level refuge requirements are important even for a highly mobile pest such as CBW. Once established, resistant CBW could spread to farms in a regions that do not use

Bt. Preliminary modeling work conducted by Storer showed that soybean, as an alternate host for CBW, can slow the rate at which resistance evolves in eastern North Carolina, where soybean is planted on roughly the same number of hectares as corn and cotton combined.

How Models Were Used to Aid the Regulatory Decision

In deciding on the size, proximity, configuration, and care of the non-Bt cotton refuge, EPA has taken into account a number of factors. EPA has used models developed to predict the estimated time that resistance would develop to compare IRM strategies for Bt crops. Because these predictive models cannot be validated without actual field resistance, they have limitations. Such models can only be used as a part of the weight of evidence determination conducted by EPA to assess the risks of resistance developing in target pest populations. Models are an important tool in determining appropriate Bt crop IRM strategies. However, as the SAP noted, model design should be peer reviewed and parameters validated (SAP 2001). In the absence of field resistance, models are "the only scientifically rigorous way to integrate all of the biological information available, and that without these models, the Agency would have little scientific basis for choosing among alternative resistance management options" (SAP 2001). While the absolute number of years to resistance is not precisely determined from the models, the relative difference in effectiveness between refuge options can be anticipated. Thus the utility of the models is not that they make accurate quantitative predictions, rather, they enable informed assessments of various refuge options.

As discussed above, EPA used five models in its comparative evaluation of refuge options. Each of these models has limitations based on the assumptions in the models. For example, the predictions generated by the models are very sensitive to assumptions about the genetics of resistance (gene frequency and functional dominance) about which little is known. EPA recognized the relative uncertainty in the predicted outcomes of these models. The predictive reliability of the models increases as other factors such as level of Bt crop adoption, level of grower IRM compliance, fitness costs of resistance to the insect, presence and availability of alternate insect host plants, spatial components, stochasticity, and pest population dynamics are better understood and included in the models. Such parameters, however, serve to increase the reliability of the modeled predictions only to the extent that the inputs reflect reality.

Certain models assume 100% compliance and 100% adoption of Bt cotton. Both of these assumptions are not realistic based on data evaluated by the Agency. One hundred percent adoption has not occurred, and therefore if models were able to address this factor, they might lengthen time-to-resistance with less than 100% adoption. Although compliance with IRM requirements by cotton growers is generally high, some growers do not fulfill every requirement; this difference would likely shorten time-to-resistance. It is not known how these two factors in combination will impact predictions of time-to-resistance or modeling uncertainty. With respect to incorporation of a parameter addressing alternate hosts, as discussed in the Agency's risk assessment document, two SAPs (1998 and 2000) stated explicitly that alternate hosts for TBW and CBW could not be considered as a refuge until there were empirical data to support their inclusion.

For alternative hosts to be effective, there must be synchronous emergence of reproductively active susceptible insects from the refuge and potentially resistant insects from Bt cotton acreage. In addition, the alternative host plants must be close enough for such susceptible insects to mate with the potentially resistant insects in the Bt cotton fields. Unfortunately, the available data are not sufficient on the biological equivalence of the insects produced on the various host plants to evaluate the timing or spatial issues. Similarly, EPA does not have adequate information on the size and proximity of such potential alternative host acreage to Bt cotton fields to evaluate how likely insects from the alternative hosts would be to mate with potentially resistant insects. Without adequate information, there is no basis to rely upon alternate hosts to provide suitable numbers of susceptible TBW or CBW; therefore, only non-Bt cotton may be used as a refuge. However, if alternate hosts can be empirically validated to function as a refuge, the models may predict longer time-to-resistance for an IRM approach.

Given the uncertainty of predictive models, EPA required that the registrant provide additional data to evaluate whether other factors, such as, alternate hosts, level of compliance, and level of adoption alter the predictions of the models. Until such time as these additional data become available, some of the models predictions may be overly conservative. However, given that EPA considers the development of resistance to be a significant adverse effect, it is prudent for the Agency to err on the side of conservative regulatory practice.

The current data and the insect resistance management models indicate that 5% external, unsprayed refuge option has a significantly greater likelihood of insect resistance than either the 5% embedded or

20% external, sprayed refuge options. The 2000 SAP stated that the external, unsprayed option poses the highest risk to resistance evolution, especially for CBW. Therefore, the external, unsprayed option expires after three growing seasons (September 30, 2004). Leading up to this, the registrant is required to develop considerable new data on alternative host plants as possible effective refuges, and study the impact of chemical insecticide sprays on Bt cotton. If, based upon these and any other pertinent data, the registrant requests extension of the expiration date for the external, unsprayed option, EPA will conduct a comprehensive assessment of whether all relevant data support such regulatory action, as part of a larger requirement that would also likely involve alternative host plants.

STRUCTURED REFUGE REQUIREMENTS

EPA has mandated that the registrant ensure that all producers of Bt cotton be required to employ one of the following structured refuge options. For the first three options, the variety of cotton planted in the refuge must be comparable to Bt cotton, especially in the maturity date, and the refuge must be managed (e.g., planting time, use of fertilizer, weed control, irrigation, termination, and management of other pests) similarly to Bt cotton.

External, Unsprayed Refuge

The refuge must be at least 5 acres of non-Bt cotton (refuge cotton) planted for every 95 acres of Bt cotton. The size of the refuge must be at least 150 feet wide, but preferably 300 feet wide. This refuge may not be treated with sterile insects, pheromones, or any insecticide (except listed below) labeled for the control of TBW, CBW, or PBW. The refuge may be treated with acephate or methyl parathion at rates that will not control TBW or the CBW (equal to or less than 0.5 lbs active ingredient per acre). The non-Bt cotton refuge must be maintained within at least 1/2 linear mile (preferably adjacent to or within 1/4 mile or closer) from the Bt cotton fields. This option expires after the 2004 growing season unless extended by amendment: EPA intends to review the data specified in the data requirements concerning alternate hosts and chemical insecticide sprays applied to Bt cotton, and decide in 2004 whether the new data support continuation of an external, unsprayed refuge as part of a larger requirement that would also likely involve alternative host plants. If these data support the continued availability of the external, unsprayed refuge option, EPA may maintain the availability of this option.

External Sprayed Refuge

The refuge must be at least 20 acres of non-Bt cotton planted as a refuge for every 80 acres of Bt cotton (total of 100A). The non-Bt cotton may be treated with sterile insects, insecticides (excluding foliar Bt products), or pheromones labeled for control of the TBW, CBW, or PBW. The refuge must be planted at least within 1 linear mile (preferably within 1/2 mile or closer) of the Bt cotton fields.

Embedded Refuge

The refuge must be at least 5 acres of non-Bt cotton (refuge cotton) planted for every 95 acres of Bt cotton. The non-Bt cotton refuge must be embedded as a contiguous block surrounded on all sides by the Bt cotton field, but not at one edge of the field. For very large fields, multiple blocks across the field may be used. For small or irregularly shaped fields, neighboring fields farmed by the same grower can be grouped into blocks to represent a larger field unit, provided the block exists within one square mile of the Bt cotton, and the block is at least 150 feet wide, but preferably 300 feet wide. Within the larger field unit, one of the smaller fields planted to non-Bt cotton may be utilized as the embedded refuge. This refuge may be treated with sterile insects, any insecticide (excluding foliar *Btk* products), or pheromones labeled for the control of TBW, CBW, or PBW whenever the entire field is treated. The refuge may not be treated independently of the surrounding Bt cotton field in which it is embedded (or fields within a field unit).

Embedded Refuge for Pink Bollworm Only

The non-Bt refuge cotton planted must be at least one single non-Bt cotton row for every six to ten rows of Bt cotton. The refuge may be treated with sterile insects, any insecticide (excluding foliar *Btk* products), or pheromones labeled for the control of PBW whenever the entire field is treated. The in-field refuge rows may not be treated independently of the surrounding Bt cotton field in which it is embedded. The refuge must be managed (fertilizer, weed control, etc.) identically to the Bt cotton. There is no field unit option as described in Embedded Refuge above.

Optional Community Refuge Pilot

This option allows multiple growers to manage refuge for external, unsprayed, and external, sprayed refuge options or both. This option is not allowed for the embedded/in-field options. A community refuge program was allowed as a continuing pilot for the 2002 growing season. EPA will evaluate the community refuge program following the 2002 growing season. The community refuge for insect resistance management must meet the requirements of either the 5% external unsprayed refuge and/or the 20% sprayed option, or an appropriate combination of the two options. The registrant has provided an explanation of the 2002 community refuge pilot program in the Bollgard Cotton 2002 Refuge Guide. Copies of field maps must be maintained by the community refuge coordinator. The registrant must annually report to the EPA on the effectiveness of the program that includes a phone audit and on-farm visits.

RESISTANCE MONITORING REQUIREMENTS

Proactive resistance detection and monitoring is critical to the survival of Bt technology. Bt cotton registrants are required to monitor for insect resistance (shifts in the frequency of resistance-conferring alleles) to Bt insecticides as an important early warning sign to resistance development in the field and to determine whether IRM strategies are working. An additional value of resistance monitoring is that it may provide potential validation of parameters used in IRM models. Effective monitoring programs should have well-established baseline susceptibility data, sensitive detection methods, and a reliable collection network. Chances of finding resistant larvae in Bt cotton depend on level of pest pressure, frequency of resistant individuals, number of samples, and sensitivity of the detection technique. Therefore, as the frequency of resistant individuals or the number of collected samples increases, the likelihood of sampling a resistant individual increases (Roush and Miller 1986). The goal is to detect resistance in an insect population before the occurrence of widespread crop failures, and if possible, in time so that mitigation practices can delay or counteract the development of resistance.

Monitoring for resistance should be undertaken in areas where the pests are known to regularly overwinter. Other secondary pests also may need to be monitored (on an individual basis), because these pests may be of local or regional significance. Previous experience with conventional insecticides has shown that once resistant phenotypes are detected at a frequency > 10%, control or crop failures are common (Roush and Miller 1986). Because of sampling and sensitivity limitations, resistance could develop to Bt toxins before it is easily detected in the field. Sampling locations should be selected to reflect all crop production practices and should be separated by a sufficient distance to reflect distinct populations but should focus on intensively planted Bt crop areas in which selection pressure is expected to be higher.

Resistance detection and monitoring can be difficult and imprecise because rare genes are hard to detect. For example, if the phenotypic frequency of resistance is 1 in 1,000 then > 3,000 individuals must be sampled to have a 95% probability of detecting one resistant individual (Roush and Miller 1986). Several methods have been proposed: (1) grower reports of unexpected damage, (2) systematic field surveying of Bt corn, (3) discriminating concentration assay, (4) F_2 screen, (5) screening against resistant colonies, and (6) sentinel Bt crop field plots. Each of these methods is discussed in detail in EPA's Bt Plant-Incorporated Protectants Biopesticides Registration Action Document (BRAD) (USEPA 2001).

In addition, DNA markers may also be useful as a screening tool. In a first step toward more efficient DNA-based monitoring, Gahan et al. (2001) describe using a DNA-based screening system for detecting Cry1Ac-resistant TBW that have developed resistance through a specific mutation in the cadherin gene (the mechanism of Bt resistance found in the YHD2 strain, see Gould et al. 1997). This mutation results in a truncated cadherin that lacks the toxin-binding region and thus cannot bind Cry1Ac. The power of DNA-based screening depends on the diversity of resistance-conferred mutations. Field populations of TBW might harbor this same mutation, other mutations of the same gene, or other genes and mechanisms of resistance. The Gahan et al. findings are the first to identify a DNA-based screening for Bt-resistant TBW heterozygotes by directly detecting the recessive allele. Their marker is being evaluated in the field, and other DNA markers are being screened.

After five years of analyzing resistance monitoring data (1996-2000), there is no evidence of TBW, CBW, or PBW resistance to Cry1Ac produced by Bt cotton (see discussion in USEPA 2001). At this time, the Agency believes that these empirical data substantiate the success of the external unsprayed, external sprayed, and embedded refuge options to delay resistance. In addition, the Agency is mandating additional improvements to the current resistance management programs that will improve the detection of resistance. The current resistance monitoring program for TBW, CBW, and PBW is summarized below.

Current Monitoring Practices

EPA has imposed specific monitoring requirements on the registrant of Bt cotton products (USEPA 2001). EPA has mandated monitoring for resistance and/or trends in increased tolerance in TBW, CBW, and PBW. There were approximately 5.8 million acres of Bollgard Bt cotton planted in the 2001 growing season and 4.5 million acres planted in the 2000 growing season (Monsanto 2002, USEPA 2001, Table 1). It would be logistically and practically impossible to sample every farm that planted Bollgard Bt cotton. Therefore, current resistance monitoring programs have focused sampling in areas of highest adoption of the Bt crops as the areas in which resistance risk is greatest.

For TBW and CBW, over 20 specific collection sites have been established for the 2003 growing season. Sites are focused in areas with high risk of resistance (e.g., where adoption is at least 75% of the cotton planted in that county or parish) while overall being distributed throughout the areas where TBW and CBW are important pests with a goal of having sites in AL, LA, AR, MS, FL, VA, GA, NC, SC, TN, and TX. For PBW, collection sites are focused in areas of high adoption, with the goal of including all states where PBW is an economically important pest (i.e., AZ, CA, NM, TX). There is a sampling goal stipulated to collect at least 250 individuals from any one location with a target of least 20 locations for TBW, CBW, and PBW. The greater the number of samples and locations, the greater the probability that any emerging resistant individuals will be detected.

The currently required, basic detection method has been a discriminating dose/diagnostic dose bioassay system that would distinguish between resistant and susceptible phenotypes. But such tests have been criticized as being too insensitive to provide early detection before resistance develops or can spread very far, especially if the alleles for resistance are rare in the insect population. Discriminating dose bioassays are most useful when resistance is common or conferred by a dominant allele (resistance allele frequency > 0.01) (Andow and Alstad 1998, Andow et al. 1998). It is currently considered as one of the central components of any monitoring plan, but other monitoring methods, such as the F_2 screen and DNA markers, may have value in conjunction with the discriminating dose assay.

The cost of the Bollgard Bt cotton resistance monitoring program is currently borne by Monsanto as the sole registrant of Bt cotton products. Researchers at the University of Arizona at Tuscon, AZ for PBW, and at the USDA/ARS research lab at Stoneville, MS for TBW and CBW are charged with carrying out the susceptibility bioassays.

Cross-Resistance

Cross-resistance occurs when a pest becomes resistant to one Bt ICP that then allows the pest to resist other, separate Bt ICPs. Efforts are underway to assess whether cotton insects show cross-resistance to various Bt proteins. Some pests of cotton are also pests of other crops for which Bt transgenic varieties or microbial Bt insecticides are available (e.g., CBW on corn, fall armyworm (*Spodoptera frugiperda* J. E. Smith) on tomato). Future resistance monitoring methods may incorporate such information because cross-resistance is an area of major concern for resistance management and poses risks to both transgenic Bt crops and microbial Bt insecticides. Cross-resistance also poses a risk to pyramid strategies, in which multiple proteins are deployed simultaneously in the same hybrid. To date, the development of cross-resistance has not been shown in insect pests exposed in the field to Bt crops producing different Bt proteins.

Discussions of cross-resistance are complicated due to the fact that the exact nature and genetics of Bt resistance are not fully understood. Resistance may vary substantially from pest to pest, adding to the unpredictability of the system. In general, it is possible for resistance to Bt proteins to occur through several different mechanisms, some of which may result in cross-resistance to other proteins. The most well documented mechanism of resistance is reduced (midgut) binding affinity to Bt proteins (reviewed in Tabashnik 1994). Different Cry proteins may bind to distinct receptors in an insect gut. Modifications to these insect gut receptors have been implicated in resistance to Cry proteins. An example of a possible shared binding site resulting in cross-resistance was observed with TBW. In this case, a laboratory strain of TBW selected for resistance to Cry1Ac was also found to be resistant to the Cry1Aa, Cry1Ab, and Cry1F proteins (Gould et al. 1995). Other mechanisms that may lead to resistance (and ultimately cross-resistance) include protease inhibition, metabolic adaptations, gut recovery, and behavioral adaptations (Heckel 1994, Tabashnik 1994).

The complexity of cross-resistance within a single species or different species is demonstrated by a wealth of experimental evidence. Examples involving TBW provide an illustration. Gould et al. (1995) selected a TBW strain (YHD2) for a high level of resistance to Cry1Ac (approximately 2000-fold). The YHD2 laboratory-selected strain was found to be cross-resistant to Cry1Aa, Cry1Ab, and Cry1F and showed limited cross-resistance to Cry1B, Cry1C, and Cry2A. Genetic experiments revealed that resistance in the YHD2 strain is partially recessive and is controlled mostly by a single locus or a set of tightly linked loci (Heckel et al. 1997). These results differ from Gould et al. (1992) using a more moderately-resistant laboratory strain of TBW (< 50-fold) which showed some broad-spectrum resistance to Cry1Aa, Cry1Ab, Cry1B, Cry1C, and Cry2A. The resistance levels in this TBW strain were low, and subsequent work showed that resistance was inherited as a nearly additive trait (Heckel et al. 1997). These results show that cross-resistance in TBW follows a variable pattern for a closely related group of proteins. Therefore, it is difficult to predict what cross-resistance patterns are likely to be in the field because evolutionary responses will depend on the initial frequencies of each resistance allele, the dominance of the alleles, and how the proteins are used.

Because of the complexity and uncertainty associated with predicting cross-resistance, the Agency has taken measures to evaluate the cross-resistance of pest species to the Cry proteins expressed in Bt plants. EPA required that registrants submit data evaluating the cross-resistance potential of various insect pests to Bt proteins prior to registration.

Insects such as the TBW have been shown to have a broad cross-resistance potential to Cry1A, Cry1F, and Cry2A proteins (Gould et al. 1992). Cross-resistance issues are relevant to current Bt crops, especially Bt corn and Bt cotton, that deploy Cry1A proteins in which TBW cross-resistance to a number of Bt proteins has been demonstrated in laboratory binding studies. Cross-resistance is also important to the livelihood of organic growers who use Bt foliar sprays on crops in which CBW is a problem.

One new Bt cotton product (Bollgard II™) has both the *cry1Ac* and *cry2Ab* genes expressed within the same cotton cultivars. Insect resistance management strategies need to account for both Cry1Ac and Cry2Ab being pyramided in Bt (Bollgard II) cotton cultivars. Based on the available literature examining the receptor binding properties of Cry1A and Cry2A in CBW, TBW, and European corn borer (*Ostrinia nubilalis* Hübner) larvae, it is not likely that high levels of cross-resistance would develop to Cry2A if resistance develops to Cry1A in commercially available Bt corn and Bt cotton. Based on the work of English et al. (1994), Cry1A and Cry2A proteins exhibit different binding characteristics and possess different modes of action.

Work with TBW and CBW resistant (to Cry1Ac) colonies indicates that there is some low potential for cross-resistance, and that there are likely to be a range of Bt resistance mechanisms (Bradley et al. 2001, Gould 2001). Previously published research indicates that there is evidence for broad cross-resistance at low levels to Cry1A and Cry2A in laboratory-selected strains of beet armyworm (Moar et al. 1995) and TBW (Gould et al. 1992). Preliminary bioassays conducted on PBW by Dennehy et al. (2001) showed that resistance to Cry1Ac in a resistant PBW strain (AZP-R) does not confer cross-resistance to Cry2Ab.

REMEDIAL ACTION PLANS

EPA required a remedial action plan be available in the unfortunate situation that resistance is suspected or actually does develop. Again, as for resistance monitoring plans, remedial action plans are specific for the crop and pest. For example, because the PBW is predominately a pest of cotton in the western US and has such a different biology than the other two target pests of Bt cotton, the remedial action plan for PBW is quite different than that for either CBW and/or TBW in the southeastern US. These plans define not only suspected and confirmed resistance, but also the key steps and actions needed if and when resistance develops. Generally, if resistance is confirmed, the farmers involved will treat their Bt crop with alternative pest control measures. This might be a chemical pesticide known to be highly effective against the insect or it might mean measures such as crop destruction. In addition, the sales and distribution of the Bt crop would be suspended in that area and the surrounding area until it can be determined that insects in that area have regained their susceptibility to the Bt product. There would also need to be increased monitoring to define the remedial action area(s). Other remedial action strategies include increasing refuge size, changing dispersal properties, use of sterile of insects, or use of other modes of actions. Geospatial surveys would help define the scale of remedial action and where to intensify monitoring.

Because no field resistance has yet been found to any of the Bt crops, all of these tactics are untested. However, a key attribute of these plans is having farmer involvement in plan development. So far there is only a regional remedial action plan for the Arizona area where PBW is the chief pest controlled by Bt cotton. The Arizona Bt Cotton Working

Group has produced "A Remedial Action Plan for PBW Resistance to Bt Cotton in Arizona" which is thorough and EPA believes is very workable (see Appendix 1, USEPA 2001). An interim remedial action plan is currently required and is being revised to address TBW and CBW resistance to Bt cotton, key economic pests of cotton in the mid-South and Southeastern US (see Appendix 2, USEPA 2001).

Suspected Resistance

EPA defines "suspected" resistance to mean, in the case of reported product failure, that:

- the cotton in question has been confirmed to be Bt cotton;
- the seed used had the proper percentage of cotton expressing Bt protein;
- the relevant plant tissues are expressing the expected level of Bt protein; and
- it has been ruled out that species not susceptible to the protein could be responsible for the damage, that no climatic or cultural reasons could be responsible for the damage, and that other reasonable causes for the observed product failure have been ruled out.

The Agency does not interpret "suspected resistance" to mean only grower reports of possible control failures, nor does the Agency intend that extensive field studies and testing to fully scientifically confirm insect resistance be completed before responsive measures are undertaken.

If resistance is "suspected," the registrant must instruct growers to do the following:

- Use alternate control measures to control the pest suspected of resistance to Bt cotton in the affected region.
- Destroy crop residues in the affected region immediately after harvest (i.e., within one month) with a technique appropriate for local production practices to minimize the possibility of resistant insects overwintering and contributing to the next season's pest population.

Confirmed Resistance

The registrant assumes responsibility for the implementation of resistance mitigation actions undertaken in response to the occurrence of resistance during the growing season. When resistance has been confirmed, the registrant must stop sale immediately and distribution of Bt cotton in the remedial action zone (may be less than a single county, single county, or multiple counties) where the resistance has been shown until an effective local mitigation plan approved by EPA has been implemented.

GROWER EDUCATION, ATTAINING COMPLIANCE

Education

For growers to plant refuges according to guidelines, they must be presented with consistent and up-to-date messages. Monsanto has invested in programs and materials to educate growers on the value of incorporating the IRM plan into their farming practices. Monsanto conducts numerous grower and retailer meetings and provides financial support to academic and extension researchers. Monsanto also performs annual grower compliance surveys and field visits. Specific scouting techniques have been developed for Bt cotton. A partnership developed between industry, National Cotton Council, State grower organization, universities, extension experts, consultants, and state/federal governmental regulatory agencies could be beneficial to promote insect resistance management. One example of very good partnership is the Arizona Bt Cotton Working Group.

The Agency recognizes that IRM strategies should be scientifically sound, but they must also consider the practical, logistical, and economic needs of growers (a fine balance). There is no question that IRM plans should always be scientifically valid. But, there is a question about whether growers will implement IRM plans that are not practical. High grower adoption of IRM strategies is ultimately the most important aspect of IRM. The best of plans will fail, if most growers cannot or will not use them. Generally, field specialists have taken a common-sense approach to IRM and try to work within the growers' equipment and field limitations. One approach is to offer growers a "toolbox" of options that all fulfill the IRM requirements so that they can choose IRM plans that best fit their region of the country or individual operation. Growers should be allowed some flexibility with their plans, but any flagrant consistent misuse of the Bt technology must be dealt with accordingly.

Growers are the most essential component of a successful IRM program because they are ultimately responsible for planting refuges and carrying out the details of an IRM plan. Thus, a program that educates growers about the importance of IRM and follows with compliance monitoring is an integral part of any resistance management strategy.

Compliance

In addition to carrying out effective IRM education for growers, Bt cotton registrants are required to establish a broad compliance program as part of the IRM requirements (USEPA 2001, see III. "Bt Cotton Confirmatory Data and Terms and Conditions of the Amendment"). Ideally, a compliance program should (1) establish an enforcement structure that will maximize compliance, (2) monitor level of compliance, and (3) investigate effects of noncompliance on IRM.

Grower compliance with refuge and IRM requirements is a critical element for resistance management. Significant non-compliance with IRM among growers may increase the risk of resistance to Bt. However, it is not known what level of grower non-compliance will compromise the risk protection of current refuge requirements.

The Agency recognizes that compliance is a complex issue for Bt crops and IRM. There is currently disagreement as to the appropriate refuge size/deployment and the level of grower compliance necessary to achieve risk reduction. Given that the Agency's goals are to preserve pest susceptibility to Bt because it is in the "public good" and to protect the benefits of Bt crops, IRM (and subsequently grower compliance) is very important. Optimally, refuge requirements would change over time as pest susceptibility changes. However, changes to refuge requirements are difficult to implement. Therefore, the Agency must set safe refuge requirements that preserve the pest(s) susceptibility and protect the benefits of Bt crops. Currently, the financial burden of implementing these refuge requirements is borne primarily by the growers. Compliance costs for monitoring are borne by the registrant. Increasing refuge size and/or limiting refuge deployment to better mitigate the risk of resistance is likely to increase costs to growers and result in a higher rate of grower non-compliance. Grower compliance with IRM strategies for current Bt crops is tied into the belief that future technologies, such as plants expressing multiple Bt toxins and other new PIPs, will reduce the risk of resistance.

Enforcement Structure

The first element of a system to ensure a high level of compliance is a mechanism to create a legally enforceable obligation on Bt cotton growers to comply with the refuge program. This is accomplished through "grower agreements" (USEPA 2001, see III14). Registrants have flexibility to design grower agreement programs that fit their own business practices. As part of the compliance assurance plan, each registrant must establish and publicize a "phased compliance approach." That is, a guidance document that indicates how the registrant will address noncompliance within the terms of the IRM program and general criteria for choosing among options for responding to noncompliant growers (USEPA 2001, see III14-17). There must be a consistent set of standards for responding to noncompliance. The options will include withdrawal of the right to purchase Bt cotton for an individual grower or for all growers in a specific region. An individual grower found to be significantly out of compliance in two successive years would be denied sales of the product the next year. Similarly, seed dealers who are not fulfilling their obligations to inform or educate growers of their IRM obligations might lose their opportunity to sell Bt cotton if their violations are proven to be significant.

Compliance Monitoring

Beginning with the 2002 growing season, the IRM compliance assurance program includes an annual survey of a statistically representative sample of Bt cotton growers conducted by an independent third party. The survey should measure the degree of compliance with the IRM program by growers in different regions of the country and consider the potential impact of nonresponse. The registrant is required to provide EPA with compliance monitoring report and plans for updating the compliance assurance plan on an annual basis. The registrant is also required to follow up on "tips and complaints" concerning noncompliance with IRM requirements.

Monsanto and its representatives have been conducting grower surveys and on-farm visits since 1996 (Table 5). Monsanto representatives looked at field maps, visited fields, and used the gene check kits to confirm the refuge cotton plants were not-Bollgard. The IPM practices discussed included: scouting followed by selective insecticide use to enhance natural enemy populations for additional control; managing for early maturity of varieties; post-harvest stalk destruction to minimize resistance to Bollgard in late-season infestations; and soil management

TABLE 5. Percent Grower Compliance–Monsanto Study

Year	% Growers Following the Refuge Guidelines
1996	99
1997	98
1998	91
1999	94
2000	95

practices that encourage destruction of over-wintering pupae. Explicit distance requirements for each refuge option were not mandated by EPA until the 2001 growing season. Therefore, these results indicate just the percentage of growers who followed the refuge size requirements.

In 2001, there were 144 community refuge pilots (consisting of 166,000 acres of Bollgard cotton). The primary refuge option used was the 20% external, sprayed structured refuge. Arkansas (43) and Mississippi (27) had the most community refuges. Texas and Georgia had sixteen community refuges each and North Carolina had fourteen. The number of participants per community ranged from two to six with two being the most common (98 of 144 total communities). Telephone surveys and on-farm visits were conducted by Monsanto to assess compliance with the refuge requirements. Of those surveyed, all met the appropriate refuge size and distance requirements. The community refuge program was very successful in accomplishing its goal in 2001. It was useful in helping growers to meet the distance requirements that were mandated as part of the refuge requirements in 2001.

RESEARCH DATA REQUIREMENTS

The Agency has made the determination that some additional IRM data are needed to characterize better the impact of alternate hosts and supplemental insecticide treatments on refuge effectiveness, and north-south movement of CBW.

Alternate Hosts as an Effective Refuge and Insecticide Sprays on Bt Cotton

The SAP meetings held in 1998 and 2000 concluded that only non-Bt cotton could be used as an effective refuge until more data are gathered regarding alternate hosts as effective refuges. While soybeans, sorghum, and wild hosts may serve as hosts for these insects (see discussion in USEPA 2001), explicit empirical data are not yet available to ensure that an effective refuge is present. The needed data will address the timing of adult insect emergence and distribution of insects on each alternate host, density, fitness of adults emerging from other crops, and spatial arrangement of the planting areas for other crops. EPA has mandated that Monsanto conduct research on the effectiveness of different alternate hosts as suitable refuge for CBW. Research topics must include, but are not be limited to: (1) mating and oviposition behavior of CBW (*H. zea*), (2) fitness of adults and adult population densities coming from the alternate hosts vs. unsprayed and sprayed Bt cotton, (3) whether insect pest emergence is in synchrony with pests emerging from Bt cotton, (4) proximity of alternate hosts to Bt cotton, and (5) design of new resistance management models that include alternate hosts appropriate for different cotton production regions (e.g., North Carolina vs. Louisiana). Studies must be conducted across the cotton belt where CBW is an economic pest. The sites must represent a range of conditions that will affect CBW biology. Conditions must include, such factors as irrigation, soil types, and climatic conditions.

In addition, study is required on whether Bt-resistant CBW would survive supplemental insecticidal treatments and increase the potential effectiveness of non-Bt refuges. Research studies must be conducted to determine the IRM value of different insecticide chemistries likely to be used against the CBW in conventional and transgenic Bt cotton (irrigated and non-irrigated, side by side field trials). Any potential effects must be related to survival of putative Bt-resistant CBW and effective refuge size. Usage data must be provided for insecticide use on Bt cotton fields from 1997 to 2001.

Once these data are collected, EPA has mandated that Monsanto must construct or update resistance management models, as appropriate, for cotton producing areas in the U.S. (i.e., areas where CBW (*H. zea*) typically exceeds economic threshold on Bt cotton) to include alternate hosts. Resistance management models must also include consideration of supplemental insecticidal treatments for control of CBW.

Reverse Migration (North/South Movement) of CBW

The Agency has mandated that Monsanto conduct field experiments on north-south movement of CBW from corn-growing regions to cotton-growing regions using C^{14} isotope decay or other suitable methods.

CBW can have several generations per year and frequently the insect moves from corn to cotton. There is not a high dose of Cry1Ac for control of CBW in either Bt corn or Bt cotton. If CBW survives exposure to Bt corn and then moves to Bt cotton, then the chances of resistance development are increased through the added exposure.

The SAP (SAP 2001) indicated that there was more evidence of CBW migration from the north to the south than evidence against this migration pattern. The Panel went on to discuss how the movement of CBW from the north to the south could impact insect resistance management, specifically refuge size. The Panel noted that as long as the amount of Bt corn in a (northern) region did not exceed 50%, then the refuge size was adequate. However, there are several areas in the Corn Belt where market penetration of Bt corn exceeds 50%.

CONCLUSION

The EPA has required an unprecedented IRM program for Bt crops. The specific IRM strategies and requirements for Bt cotton in the United States have been developed by a coalition of stakeholders including EPA, USDA, academic researchers, industry, cotton producers and producer organizations, public interest groups, and growers. Many of these stakeholders recognize that IRM strategies need to be scientifically sound, practical, flexible, implementable, and sustainable.

The current registration for Cry1Ac PIP as expressed in cotton (Bollgard cotton) expires September 30, 2006, except for the 5% external, unsprayed structured refuge option, which will expire September 30, 2004. There are mandatory refuge requirements that consist of three structured non-Bt cotton refuge options with specific deployment requirements: (1) 5% external, unsprayed refuge option, (2) 20% external, sprayed refuge option, and (3) 5% embedded refuge option. In-field refuges consisting of at least one single non-Bt cotton row for every six to ten rows of Bt cotton are allowed for PBW control only. A 5% embedded refuge is allowed for TBW and CBW management. A community refuge pilot that allows multiple growers to work together to comply with the refuge requirements was renewed for 2002. The 5% external, unsprayed and/or 20% external, sprayed refuge options may be used in the community refuge program. There are also requirements for annual resistance monitoring, remedial action plan, grower education, grower compliance, research, and annual reporting. Additional IRM research on the effect of north-south movement by CBW, effectiveness of alternate hosts as refuge, and effectiveness of supplemental insecticide sprays will allow current IRM strategies to be further improved for greater long-term sustainability.

The Agency used multiple IRM predictive models as tools to compare refuge options. Models are imperfect, but in the absence of field resistance, are the only tools available that integrate all of the available biological and genetic information to predict the likelihood of resistance. Further data are needed to validate input parameters, e.g., alternate host contribution to effective refuge size, level of adoption, level of compliance, and supplemental insecticidal spray effects. Uncertainty in models also indicates that focus is essential on resistance monitoring, grower actions, and potential remedial action strategies.

An annual resistance monitoring has been implemented for Bt cotton since it was first commercialized in 1995. This program has been proactive and has focused on areas of highest risk of resistance. However, sampling and detection methodologies need improvement. Insect resistance monitoring is expensive and the extremely high costs can be offset by more reliance on farmer actions to carry out robust IRM plans, on compliance monitoring, and thorough remedial action plans.

The science of insect resistance management, including models, is complex and is continuing to develop. Maintaining the Bollgard cotton IRM program, or any, IRM program requires the effective participation of farmers, pesticide companies, researchers, and government regulators. EPA will continue to monitor all of these activities closely for Bt cotton products and make further IRM requirements if necessary.

REFERENCES

Andow, D.A. and D.N. Alstad, (1998) The F_2 screen for rare resistance alleles. *J. Econ. Entomol.* 91: 572-578.

Andow, D.A., D.N. Alstad, Y.-H. Pang, P.C. Bolin, and W.D. Hutchison, (1998) Using a F_2 screen to search for resistance alleles to *Bacillus thuringiensis* toxin in European corn borer (Lepidoptera: Crambidae). *J. of Econ. Entomol.* 91: 579-584.

Andow, D.A. and W.D. Hutchison, (1998) Bt corn resistance management. In M. Mellon and J. Rissler (eds.), Now or Never: Serious New Plans to Save a Natural Pest Control, Union of Concerned Scientists, Two Brattle Square, Cambridge, MA.

Andow, D.A., D.M. Olson, R.L. Hellmich, D.N. Alstad, and W.D. Hutchison, (2000) Frequency of resistance to *Bacillus thuringiensis* toxin Cry1Ab in the Iowa population of European corn borer (Lepidoptera: Crambidae). *J. Econ. Entomol.* 93: 26-30.

Bourguet, D., A. Genissel, and M. Raymond, (2000) Insecticide resistance and dominance levels. *Journal of Economic Entomology*. 93: 1588-1595.

Burd, A.D., J.R. Bradley, Jr., J.W. Van Duyn, F. Gould, and W. Moar, (2001) Estimated frequency of non-recessive B.T. resistance genes in Bollworm, *Helicoverpa zea*. Proceeding of the Beltwide Cotton Conferences, National Cotton Council. pp. 820-822.

Bradley, J.R., Jr., J. Van Duyn, and R. Jackson, (2001) Greenhouse evaluation of Bollgard II cotton genotypes for efficacy against feral and Cry1Ac selected strains of *Helicoverpa zea*. Project Code: CT99MON7 and CT99MON8. Unpublished study. Monsanto submission to USEPA, Appendix 2, November 20, 2001 MRID# 455457-01.

Burd, A.D., J.R. Bradley, Jr., J.W. Van Duyn, and F. Gould, (2001) Resistance of bollworm to Cry1Ac toxin (MVP). 2000 Proceeding of the Beltwide Cotton Conferences., National Cotton Council. pp. 923-926.

Caprio, M., (2000a) Dispersal and ovipositional patterns of tobacco bollworm (*Helicoverpa zea*) in the corn-cotton agroecosystem. Unpublished study.

Caprio, M., (2000b) Personal communication to S. Matten, OPP/BPPD. E-mail dated 5/10/2000.

Caprio, M., (2000c) Personal communication to Dr. S. Matten, OPP/BPPD. E-mail dated 6/16/2000.

Caprio, M., (2001) Source-sink dynamics between transgenic and non-transgenic habitats and their role in the evolution of resistance. *J. Econ. Entomol.* 94: 698-705.

Caprio, M. and J. Benedict, (1996) Biology of the Major Lepidopteran Pests of Cotton (June 24, 1996. Unpublished data. Submission to USEPA by Monsanto, MRID 2204225-01.

Carrière, Y., T.J. Dennehy, B. Pedersen, S. Haller, C. Ellers-Kirk, L. Antilla, Y.-B. Liu, E. Willott, and B.E. Tabashnik, (2001) Large-scale management of insect resistance to transgenic cotton in Arizona: can transgenic insecticidal crops be sustained? *J. Econ. Entomol.* 94(2): 315-325.

Dennehy, T., M. Sims, K. Larkin, G. Head, W. Moar, Y. Carrière, and B. Tabashnik, (2001) Control of resistant pink bollworm by second generation Bt cotton. Unpublished study. Monsanto submission to USEPA, Appendix 3, November 20, 2001. MRID# 455457-01.

English, L., H.L. Robbins, M.A. Von Tersch, C.A. Kulesza, D. Ave, D. Coyle, C.S. Jany, and S.L. Slatin, (1994) Mode of action of CryIIA: a *Bacillus thuringiensis* delta-endotoxin. *Insect Biochem. Molec. Biol.* 24: 1025-1035.

Gahan, L.J., F. Gould, and D.G. Heckel, (2001) Identification of a gene associated with Bt resistance in *Heliothis virescens*. *Science* 293: 857-860.

Gould, F., (1998a) Evolutionary biology and genetically engineered crops. *BioScience* 38: 26-33.

Gould, F., (1998b) Sustainability of transgenic insecticidal cultivars: Integrating pest genetics and ecology. *Annu. Rev. Entomol.* 43: 701-726.

Gould, F., (2000) Personal communication to S. Matten, OPP/BPPD. E-mail dated January 31, 2000.

Gould, F., (2001) Efficacy of new Bt-cotton varieties against resistant strains of *Heliothis virescens*. Unpublished study. Monsanto submission to USEPA, Appendix 1, November 20, 2001. MRID# 455457-01.

Gould, F., A. Anderson, A. Jones, D. Sumerford, D. Heckel, J. Lopez, S. Micinski, R. Leonard, and M. Laster, (1997) Initial frequency of alleles for resistance to *Bacillus thuringiensis* toxins in field populations of *Heliothis virescens*. *Proc. Natl. Acad. Sci. USA* 94: 3519-3523.

Gould, F., A. Anderson, A. Reynolds, L. Bumgarner, and W. Moar, (1995) Selection and genetic analysis of a *Heliothis virescens* (Lepidoptera: Noctuidae) strain with high levels of resistance to *Bacillus thuringiensis* toxins. *J. Econ. Entomol.* 88: 1545-59.

Gould, F., N. Blair, M. Reid, T.L. Rennie, J. Lopez, and S. Micinski, (2002) *Bacillus thuringiensis*-toxin resistance management: stable isotope assessment of alternate hose use by *Helicoverpa zea*. *PNAS* 99: 16581-16586.

Gould, F., A. Martinez-Ramirez, A. Anderson, J. Ferré, F. Silva, and W. Moar, (1992) Broad spectrum Bt resistance. *Proc. Natl. Acad. Sci. USA* 89: 1545-1559.

Gould, F. and B. Tabashnik, (1998) Bt-cotton resistance management. In M. Mellon and J. Rissler (eds.), Now or Never: Serious New Plans to Save a Natural Pest Control, Union of Concerned Scientists, Two Brattle Square, Cambridge, MA.

Hardee, D.D., J.W. Van Duyn, M.B. Layton, and R.D. Bagwell, (2000) Bt Cotton & Management of the Tobacco Budworm Complex. U.S. Department of Agriculture, Agricultural Research Service, ARS-154. 40 pp.

Heckel, D.G., (1994) The complex genetic basis of resistance to *Bacillus thuringiensis* toxin in insects. *Biocontrol Sci. and Tech.* 4: 405-417.

Heckel, D.G., L.C. Gahan, F. Gould, A. Anderson, (1997) Identification of a linkage group with a major effect on resistance to *Bacillus thuringiensis* Cry1Ac endotoxin in tobacco budworm (Lepidoptera: Noctuidae), *J. Econ. Entomol.* 90: 75-86.

International Life Science Institute, (1999) An evaluation of insect resistance management in Bt field corn: a science-based framework for risk assessment and risk management. Report of an expert panel. November 23, 1998.

Liu, Y.-B., B.E. Tabashnik, T.J. Dennehy, A.L. Patin, and A.C. Bartlett, (1999) Development time and resistance to *Bt* crops. *Nature.* 100: 519.

Livingston, M.J., G.A. Carlson, and P.L. Fackler, (2002) Use of mathematical models to estimate characteristics of pyrethroid resistance evolution in tobacco budworm and bollworm (Lepidoptera: Noctuidae) field populations. *Journal of Econ. Ent.* 95: 1008-1017.

Mahaffey, J.S., J.S. Bacheler, J.R. Bradley, Jr., and J.W. Van Duyn, (1994) Performance of Monsanto's transgenic B.t. cotton against high populations of lepidopterous pests in North Carolina, 1994 Proceedings Beltwide Cotton Conferences. National Cotton Council. pp. 1061-1063.

Mellon. M. and J. Rissler, (1998) Now or Never: Serious New Plans to Save a Natural Pest Control. Ed. M. Mellon and J. Rissler (eds.). Union of Concerned Scientists. Cambridge, Massachusetts. pp. 149.

Moar, W.J., M. Pusztai-Carey, H. Van Faasen, D. Bosch, R. Frutos, C. Rang, K. Luo, and M.J. Adang, (1995) Development of *Bacillus thuringiensis* CryIC resistance by *Spodoptera exigua* (Hübner) (Lepidoptera: Noctuidae). *Applied and Environmental Microbiology.* 61: 2086-2092.

Monsanto Company, (2002) 2001 Bollgard™ Bt Cotton Sales Data. Unpublished study. Submission to USEPA dated January 30, 2002.

Onstad, D.W. and F. Gould, (1998) Modeling the dynamics of adaptation to transgenic maize by European corn borer (Lepidoptera: Pyralidae). *Journal of Economic Entomology*. 91(3): 585-593.

Peck, S., F. Gould, and S. Ellner, (1999) Spread of resistance in spatially extended regions of transgenic cotton: implications for management of *Helitohis virescens* (Lepidoptera: Noctuidae). *J. Econ. Entomol*. 91(6): 1-16.

Roush, R.T., (1994) Managing pests and their resistance to *Bacillus thuringiensis*: can transgenic crops be better than sprays? *Biocontrol Science and Technology*. 4: 501-516.

Roush, R.T., (1997a) Managing resistance to transgenic crops. In Advances in Insect Control: The Role of Transgenic Plants. (Ed. Carozzi, N. And Koziel, M.) pp. 271-294 (Taylor and Francis, London; 1997).

Roush, R.T., (1997b) *Bt*-transgenic crops: just another pretty insecticide or a chance for a new start in resistance management? *Pesticide Sci*. 61: 328-334.

Roush, R.T. and G.L. Miller, (1986) Considerations for design of insecticide resistance monitoring programs. *J. Econ. Entom*. 79: 293-298.

Scientific Advisory Panel (SAP), Subpanel on *Bacillus thuringiensis* (Bt) Plant-Pesticides (February 9-10, 1998), (1998) Transmittal of the final report of the FIFRA Scientific Advisory Panel Subpanel on *Bacillus thuringiensis* (Bt) Plant-Pesticides and Resistance Management, Meeting held on February 9-10, 1998. Report dated, April 28, 1998. (Docket Number: OPPTS-00231).

Scientific Advisory Panel (SAP), Subpanel on Insect Resistance Management (October 18-20, 2000), 2001. Report: sets of scientific issues being considered by the Environmental Protection Agency regarding: *Bt* plant-pesticides risk and benefit assessments. Report dated, March 12, 2001. pp. 5-33.

Shelton, A.M. and R.T. Roush, (1999) False reports and the ears of men. *Nature Biotechnology*. 17(9): 832.

Stadelbacher, E.A., H.M. Graham, V.E. Harris, J.D. Lopez, J.R. Phillips, and S.H. Roach, (1986) Heliothis populations and will host plants in the Southern U.S. In Theory and Tactics of *Heliothis* Population Management. pp. 54-74. Southern Cooperative Bulletin Series. Bulletin 316.

Storer, N.P., F. Gould, and G.G. Kennedy, (1999) Evolution of regional-wide resistance in cotton bollworm to *Bt* cotton as influenced by *Bt* corn: identification of key factors through computer simulation. *1999 Proceedings of the Beltwide Cotton Conference*. 2: 952-956.

Tabashnik, B.E., (1994) Evolution of resistance to *Bacillus thuringiensis*. *Annu. Rev. Entomol*. 39: 47-79.

Tabashnik, B.E., A.L. Patin, T.J. Dennehy, Y.-B. Liu, E. Miller, and R.T. Staten, (1999) Dispersal of pink bollworm (Lepidoptera: Gelechiidae) males in transgenic cotton that produces a *Bacillus thuringiensis* toxin. *J. Econ. Entomol*. 92(4): 772-780.

U.S. Environmental Protection Agency (USEPA), (1998) The Environmental Protection Agency's White Paper on Bt Plant-Pesticide Resistance Management. U.S. EPA, Biopesticides and Pollution Prevention Division (7511C), 14 January, 1998 [EPA Publication 739-S-98-001].

U.S. Environmental Protection Agency (USEPA), (2001) Biopesticides Registration Action Document: *Bacillus thuringiensis* Plant-Incorporated Protectants (10/16/01), posted at <http://www.epa.gov/pesticides/biopesticides/reds/brad_bt_pip2.htm>.

U.S. Environmental Protection Agency and U.S. Department of Agriculture, (1999) Report of EPA/USDA Workshop on Bt Crop Resistance Management in Cotton. Memphis, Tennessee. August 26, 1999. Esther Day, ed. 80 pp. American Farmland Trust, Center for Agriculture in the Environment.

Watson, T.F., (1995) Impact of transgenic cotton on pink bollworm and other lepidopteran insects. 1995 Proceedings Beltwide Cotton Conferences. National Cotton Council. pp. 759-760.

A Rapid Assay for Gene Expression in Cotton Cells Transformed by Oncogenic *Agrobacterium* Strains

Kanniah Rajasekaran

SUMMARY. A simple expression assay for evaluation of gene constructs for input traits into cotton cells (*Gossypium hirsutum* L.) using oncogenic *Agrobacterium* strains is presented. Explants from three commercial cotton varieties, representing diverse genotypes, exhibited tumor or root formation to an equal degree in response to infection by different types of oncogenic *Agrobacterium* strains. Cotyledon explants readily developed tumors (100%) within a week and the tumors doubled in fresh weight every two weeks. *A. rhizogenes* super-rooting mutant strain MT232 was highly infective on cotyledon explants. Experiments with oncogenic strains served as a basis for development of an assay using tumor-inducing binary vectors carrying the gene to be evaluated, a truncated *Bacillus thringiensis* (Bt) *cry1Ac* insecticidal crystal protein (ICP) gene. The efficacy of the truncated ICP gene was first demonstrated using transgenic tobacco plants challenged with the Lepidopteran pest, tobacco hornworm (*Manduca sexta*). An oncogenic binary vector

Kanniah Rajasekaran is Research Biologist, USDA, ARS, Southern Regional Research Center, 1100 Robert E. Lee Boulevard, New Orleans, LA 70124 (E-mail: krajah@srrc.ars.usda.gov).

The author thanks John Grula, Rick Yenofsky and David Anderson for their help during this study and Allan Zipf and Tom Jacks for their suggestions for improvement.

[Haworth co-indexing entry note]: "A Rapid Assay for Gene Expression in Cotton Cells Transformed by Oncogenic *Agrobacterium* Strains." Rajasekaran, Kanniah. Co-published simultaneously in *Journal of New Seeds* (Food Products Press, an imprint of The Haworth Press, Inc.) Vol. 5, No. 2/3, 2003, pp. 179-192; and: *Bacillus thuringiensis: A Cornerstone of Modern Agriculture* (ed: Matthew Metz) Food Products Press, an imprint of The Haworth Press, Inc., 2003, pp. 179-192. Single or multiple copies of this article are available for a fee from The Haworth Document Delivery Service [1-800-HAWORTH, 9:00 a.m. - 5:00 p.m. (EST). E-mail address: docdelivery@haworthpress.com].

http://www.haworthpress.com/store/product.asp?sku=J153
10.1300/J153v05n02_05

containing a chimeric neomycin phosphotransferase II and the Bt ICP gene conferred antibiotic resistance and pesticidal activity against Lepidopteran larvae to tumor cells from cotyledon explants. The effective expression and pesticidal activity of the ICP gene towards control of a Lepidopteran cotton pest, tobacco budworm (*Heliothis virescens*), was demonstrated in this study using cotton tumor cells. The time needed to conduct the experiment with cotton tumor cells was about three to four months from the time of initiation, equivalent to the time needed for the tobacco model system. The rapidity of this assay is extremely useful in evaluation of gene constructs for input traits in the laboratory, especially in recalcitrant species such as cotton, where more than 15 months are needed for selection and regeneration of transgenic plants. *[Article copies available for a fee from The Haworth Document Delivery Service: 1-800-HAWORTH. E-mail address: <docdelivery@haworthpress.com> Website: <http://www.HaworthPress.com>]*

KEYWORDS. *Agrobacterium, Bacillus thuringiensis* (Bt), cotton, gene expression, insecticidal crystal protein (ICP), oncogenic, transformation

INTRODUCTION

Insect protected and herbicide-tolerant transgenic cottons are extremely popular in cotton-growing areas of the world as evidenced by the rapid increase in acreage under transgenic cotton varieties from 1.6 million acres in 1996 to 5.8 million acres in 2001 (Matten and Reynolds, this volume, p. 142, Table 1). Cotton is grown in approximately 16 million acres in the USA and about 75% of the area is under transgenic cotton varieties (USDA-NASS 2001; Wilkins et al. 2000). All the transgenic cotton under cultivation in the US has been produced via *Agrobacterium*-mediated transformation of Coker varieties. In comparison to commercially grown elite varieties, these obsolete varieties are preferred for producing transgenic lines because of ease of transformation and regeneration The transgenic traits are subsequently bred into commercial varieties by backcrossing. Regardless of the type of varieties used for the initial transformation and regeneration of fertile plants, the procedure is lengthy and time consuming with *Gossypium* species and varieties (Rajasekaran et al. 2001). Even with the easy-to-transform Coker varieties, it takes up to 16 months to produce transgenic plants. For this reason, gene constructs are often evaluated for their expression and performance against the targeted pests or weeds first using a model system, for example, *Arabidopsis* or tobacco. However, availability of a rapid, yet simple, expression assay in cotton cells would be beneficial to evaluate gene constructs prior to time-consuming selection and regeneration procedures. The present study highlights the rapidity with which cotton tumor cells can be produced using oncogenic Agrobacteria. Here we also describe a simple, rapid expression assay for evaluation of gene constructs encoding input traits such as insect protection, herbicide tolerance and disease resistance, in transgenic cotton cells generated using oncogenic strains of *Agrobacterium* spp. This concept is are demonstrated using the *cry1Ac* insecticidal crystal protein (ICP) gene from *Bacillus thuringiensis* (Bt) as an example in transgenic cotton cells after pre-testing in tobacco.

MATERIALS AND METHODS

Plant Materials

Acid delinted, fungicide treated cotton (*Gossypium hirsutum* L.) seed of Acala varieties SJ-2 and A1618, or Coker 315 were supplied by Phytogen Cottonseed Company, Corcoran, CA. For transformation of tobacco (*Nicotiana tabacum* L.), variety SR-1 was used.

Oncogenic Agrobacterium Strains and Vectors

The oncogenic *Agrobacterium* strains and their derivatives used in this study are listed in Table 1. The plasmid vector pCIB10 (Rothstein et al. 1987) was introduced into the different oncogenic host strains by the freeze-thaw transformation method (An et al. 1988). The vector pCIB10 contains the selectable marker gene, neomycin phosphotransferase II (NPT II) with a nopaline synthase promoter and terminator (*nos:neo-nos*).

A truncated ICP gene construct, containing approximately 645 amino acids of the full-length Bt *cry1Ac* gene, was made by removing the carboxy terminal portion of the gene by cleaving at the *Bcl* I restriction site at position 2090. The 1.9 kB *BamH* I/*Bcl* I fragment containing the truncated ICP gene under the control of a CaMV 35S promoter and terminator was ligated into *BamH* I-cleaved pCIB10. The plasmid was later introduced into Ach5 strain by the freeze-thaw method to produce an oncogenic binary vector for transformation of cotton or introduced

TABLE 1. *Agrobacterium* Strains Used in This Study

Bacterial strain/plasmid	Type	Source
A. tumefaciens		
Ach5	octopine	Dr. M-D. Chilton
A281	agropine	Dr. M-D. Chilton
C58	nopaline	Dr. M. Gordon
LBA4434	octopine LBA4404 + T-DNA from Ach5	Dr. R. Schilperoort
Ach5/pCIB10	oncogenic, binary	this study
A281/pCIB10	oncogenic, binary	this study
C58/pCIB10	oncogenic, binary	this study
Ach5/pCIB10	oncogenic, pCIB10 with the Bt cryIA(c) gene	this study
LBA4404/pCIB10	non-oncogenic, binary, pCIB10 with the Bt *cryI*A(c) gene	this study
LBA4404	non-oncogenic, control	Dr. R. Schilperoort
A. rhizogenes		
MT232	super-rooting mutant	Dr. G. Strobel
MT232/pCIB10	oncogenic, binary	this study

into LBA 4404 to produce a non-oncogenic binary vector for transformation of tobacco.

Agrobacterium cultures were grown overnight from glycerol stocks (500 µL) in YEB liquid medium at $26 \pm 2°C$ on a gyratory (120 rpm) shaker. The optical density (A_{600}) values were adjusted to 0.6-0.8 in MS liquid nutrient medium (Murashige and Skoog 1962) prior to use.

Plant Transformation

Transformation of tobacco (SR-1) was accomplished using the *A. tumefaciens*-mediated leaf disk transformation system (Horsch et al. 1985) as described by Rajasekaran et al. (2000). pCIB10 vector, containing the Bt *cryIAc* gene in non-oncogenic LBA 4404, was used for transformation. Thirty leaf disks were cultured on (MS) nutrient medium (Murashige and Skoog 1962) supplemented with 0.75 mg l⁻¹ 6-benzylaminopurine, 100 mg l⁻¹ kanamycin, and 200 mg l⁻¹ cefotaxime. The kanamycin-tolerant, adventitious shoot buds were transferred again to MS medium containing the same antibiotic (50 mg l⁻¹)

and only shoots that formed healthy root systems were subsequently transferred to pots containing a commercial soil mix for further evaluation in an environmentally-controlled growth chamber (28°C, 16 h day).

Transformations of cotton seedling explants were carried out according to the published procedures (Rajasekaran et al. 1996). Briefly, the seedling explants were treated with *Agrobacterium* suspension for 15-30 min, blotted dry, and then plated on 12 cm dia. filter paper (Whatman No. 1) placed on freshly made agar-solidified MS nutrient medium for cocultivation. This medium contained glucose (30 g l^{-1}) as the carbon source and no growth regulators. Cocultivation with residual bacteria was done in Percival incubators (26 ± 2°C, 16 h light, 40-60 µE $m^{-2} s^{-1}$) for 48 hours.

Following cocultivation, the explants were thoroughly washed in MS liquid medium containing cefotaxime (200 mg l^{-1}), blotted dry, and plated on freshly made MS basal medium without growth regulators. At this stage, the antibiotic cefotaxime was included in the media to control bacterial regrowth along with the selection agent-kanamycin (50 mg l^{-1}).

The inoculated explants were cultured in Percival incubators (28 ± 2°C, 16 h light, 60-90 µE $m^{-2} s^{-1}$) for 2 weeks. Distinct tumor growths were identified and transferred to MS basal medium after this initial culture. The explants treated with oncogenic, binary vector bearing Agrobacteria were subjected to a second subculture to provide more time for outgrowth under antibiotic selection pressure.

Analysis of Cotton Tumors and Transformed Tobacco Plants

Opine analysis of cotton tumors was performed according to the procedure of Chilton et al. (1984). NPT II ELISAs were carried out to determine the presence of the gene product (Firoozabady et al. 1987) in the potted tobacco plants and cotton tumor cells. Tumor samples that survived the double selection (autonomous growth and/or antibiotic resistance) procedures and were opine and/or npt II positive were included in the insect feeding expression assay. Extraction of DNA from npt II ELISA positive tobacco leaf samples was carried out according to the procedures of Bedbrook (1981). DNA was resuspended in TE and aliquots digested with *Xba* I. Samples were electrophoresed through 1% agarose, transferred to nitrocellulose and probed with a nick-translated npt II gene fragment using standard protocols (Sambrook et al. 1989).

For Southern blot analysis of tobacco transformants, DNAs, digested with *Xba* I to generate an internal fragment, were probed with an isolated ^{32}P-labelled full-length Bt *cry1Ac* gene.

Insect-Feeding Assay

Aseptically reared eggs of tobacco budworm (*Heliothis virescens*) and tobacco hornworm (*Manduca sexta*) laid on cheesecloth were obtained from the Tobacco Insect Control Laboratory at North Carolina State University, Raleigh, NC. Transgenic tobacco plants expressing the Bt construct under the control of CaMV 35S promoters in pCIB10 were used in feeding assays with the tobacco hornworm (*Manduca sexta*). Six first instar larvae were placed on the leaf surface of control and five transgenic plants and were left in an environmentally controlled chamber for 14 days.

Cotton tumor cells expressing the Bt construct were chopped into small pieces and used in an insect feeding assay. Tobacco budworm (*Heliothis virescens*) first instar larvae were obtained from hatched eggs and an average of six larvae were placed in each Petri plate containing the tumor cells. Feeding behavior and growth were observed over a six-day period. Control tumors, lacking *cry1Ac*, were grown on basal medium plus cefotaxime.

RESULTS

Tumor Induction by Oncogenic Agrobacterium Strains

The oncogenic strains listed in Table 1 were used to infect both cotyledon and hypocotyl explants of the Acala cotton varieties SJ-2 and A1618, or Coker 315. All the cotyledon explants from the three different genotypes readily developed tumors within a week after infection, and continued to grow on medium without growth hormones for more than six months (Table 2, Figure 1A). Tumorigenesis by *Agrobacterium* on cotton seedling cotyledon explants occurred with equal efficiency in all the varieties tested. The average size of tumors after two weeks of infection was > 3 mm for the strains Ach5, LBA4434 and C58. In comparison, the strain A281 produced smaller tumors (1.5 mm on an average). Solid, spherical greenish tumors on cotyledon explants were readily identifiable (Figure 1A) compared to those on hypocotyl explants because of copious swelling and callus proliferation. Cotyledon explants

TABLE 2. Tumor Induction on Cotyledon Explants of Cotton by Oncogenic *Agrobacterium* Strains. The Data Were Collected After Two Weeks of Culture

Strain	No. of explants with tumors/No. inoculated			Average No. of tumors per explant			Average size of tumors (mm)
	SJ-2	Coker 315	A1618	SJ-2	Coker 315	A1618	
No Vector	0/41	0/30	0/60	0	0	0	0
A.t. control (LBA4404)[a]	0/25	0/25	0/120	0	0	0	0
Ach5	59/59	30/30	n/d	6	4	-	3.4 ± 0.4
A281	51/51	30/30	58/58	10	6	8	1.5 ± 0.2
C58	47/51	22/30	n/d	2	1	1	3.0 ± 0.8
LBA4434	52/52	26/28	n/d	8	8	8	3.5 ± 0.8
MT232	70/70	25/25	25/25	24	8	8	hairy roots

[a]Non-oncogenic strain LBA4404 was used as *A.t.* control.
n/d–not determined

FIGURE 1. Cotyledon explants of cotton variety SJ-2 showing tumor/root growth induced by oncogenic *Agrobacterium* strains. Photographed after two weeks of culture on MS basal medium. (A) *A. tumefaciens* Ach5 and (B) *A. rhizogenes* MT232.

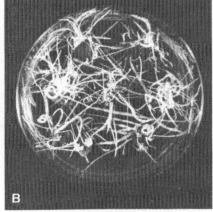

were thus utilized in subsequent experiments. Cotyledon explants also produced prolific numbers of roots (an average of 24 roots per SJ2 explant) in response to treatment with the *Agrobacterium rhizogenes* strain MT 232 (Figure 1B).

The ability of these tumors to grow on basal medium without growth hormones provided ample evidence of successful T-DNA transfer. The tumors doubled in weight approximately every 7 days and gram quantities of tumor tissue could be grown within four weeks from each event.

Selection of Antibiotic Resistant Tumors

Cotyledon and hypocotyl explants treated with oncogenic strains containing the plasmid pCIB10, which carries the marker gene that confers resistance to kanamycin, also produced tumors in equal numbers when selected first on MS basal medium and then on basal medium containing kanamycin (50 mg l^{-1}). Antibiotic-resistant (thus npt II ELISA positive) tumors were obtained at a relatively high frequency (up to 56% of the total tumors plated) from cotyledon explants (Table 3, Figure 2). The double selection (selection for autonomous growth and kanamycin resistance) procedure yielded no escapes although the growth of the antibiotic-resistant tumors was slightly slower compared to regular tumors. The doubling in fresh weight of tumor occurred every three weeks.

Antibiotic-Resistant Root Cultures

Treatment with strains of *A. rhizogenes* produced several hairy roots (Figure 1B). Isolated root terminal segments (10 mm long) were placed on MS basal medium containing 50 mg l^{-1} of kanamycin. Nearly 40%

TABLE 3. Tumor Growth Under Kanamycin Selection (50 mg l^{-1})

	Ach5/pCIB10	A281/pCIB10	C58/pCIB10
No. tumors plated	470	380	250
No. of tumors resistant to Kan 50 mg l^{-1}	214	154	140
% of antibiotic resistant tumors	46	41	56

Cotyledon explants of cv. SJ-2 were inoculated with the oncogenic strains carrying pCIB10. The data were collected after four weeks of antibiotic selection.

FIGURE 2. Antibiotic selection of tumor cells isolated from cotyledon explants of var. SJ-2 on MS basal medium plus kanamycin 50 mg l^{-1}. Tumors were first induced on MS medium and then transferred to MS medium containing kanamycin. (A) Control tumors and (B) Ach5/pCIB10 tumors on kanamycin. Photographed two weeks after transfer.

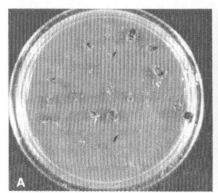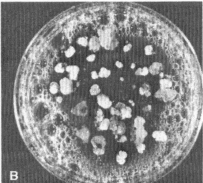

of the root tips continued to grow and elongate on antibiotic selection showing the efficient transfer of pCIB10 T-DNA (data not shown).

Analysis of Transgenic Tumors and Tobacco Plants

All antibiotic resistant cotton tumor samples and transgenic tobacco plants tested positive in the npt II ELISA. In addition, cotton tumors tested positive for the respective opines (not shown). To determine whether Bt sequences were present, DNAs isolated from transgenic tobacco plants or cotton tumor cells were digested with *Xba* I to generate a 1.8 kb internal fragment. A full-length Bt probe showed hybridization with DNA fragments of similar size (not shown).

Insect Feeding Assays with Tobacco and Cotton Tumors Expressing Cry1Ac

The transgenic tobacco plants expressing *cry1Ac*, were completely resistant to *Manduca sexta* larvae (Figure 3) compared to controls, which were defoliated by the caterpillars, thus demonstrating the efficiency, *in planta*, of the Bt construct developed for this study. The size of the larvae increased to more than 65 mm after 14 days of feeding on control plants, whereas all the larvae stopped feeding on the first day and dropped off the transgenic plants (Table 4). The insecticidal activity of Cry1Ac in cotton tumor cells was also evident. *Heliothis virescens* larvae readily consumed control tumor cells. The larvae grew rapidly in size, reaching a maximum length of 2 cm by the end of the six-day period, by which time all of the tissue provided had been eaten (Table 4). Many of the larvae placed on tumor cells expressing the Cry1Ac stopped feeding and died in three to four days. Figure 4 illustrates the

FIGURE 3. *Manduca sexta* (tobacco hornworm) feeding assay on tobacco plants. (A) Control, non-transgenic tobacco; arrow and inset, a larva 14 d after feeding. (B) Tobacco transformed with the Bt *cry1Ac* gene conferring pesticidal activity against the insect.

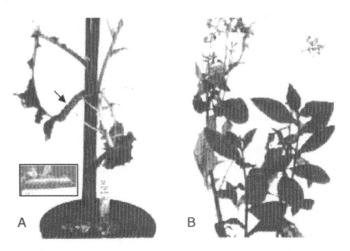

TABLE 4. Average Size of Lepidopteran Larvae Fed on Transgenic Cotton Tumors (var. SJ-2) and Tobacco Plants (var. SR-1) Expressing the Bt ICP Cry1Ac

Transgenic material	Length of tobacco budworm (mm) on cotton tumors	Length of tobacco hornworm (mm) on tobacco plants
Control, non-transformed	12 ± 1.2	65 ± 2.7
Ach5/pCIB10 tumors expressing the Cry1Ac ICP	3 ± 0.4	n/d
Tobacco plant expressing the Cry1Ac ICP	n/d	2.1 ± 0.2

The length of the caterpillars (n =15) was measured after four days of feeding on cotton tumor cells and 14 days on tobacco plants.

FIGURE 4. *Heliothis virescens* (tobacco budworm) feeding assay with cotton (var. SJ-2) tumors. (A) Kanamycin-resistant tumors from cotyledon explants treated with Ach5/pCIB10. (B) Transgenic tumors expressing Bt Cry1Ac. (C) Comparison of the sizes of larvae fed on control tumors, 'a' and kanamycin-resistant tumors, 'b' expressing the Bt ICP Cry1Ac. Photographed after four days.

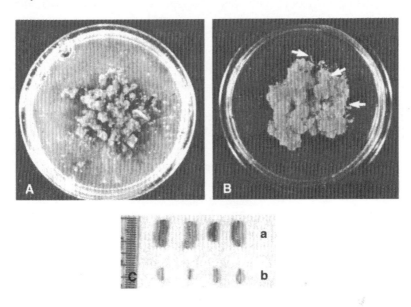

feeding behavior and the differences in larval growth on control vs. transgenic samples after four days.

DISCUSSION

It is often necessary to test several promoter:gene constructs for input or output traits prior to developing a commercially viable, safe genetically-engineered crop. For this purpose, several laboratories around the world have developed model systems such as tobacco and *Arabidopsis* because of the ease with which these species could be transformed in a relatively short time, about 1-3 months. Results obtained with model species have been useful to study the integrity and functionality of gene constructs. However, it is not necessarily possible to extrapolate their expression patterns in the targeted crop species. The long time needed in some difficult-to-regenerate crops, such as cotton, further exacerbate and delay testing. To alleviate this problem, a rapid assay for testing gene constructs for input traits in cotton cells has been developed. This was achieved by using tumor cells expressing the trait of interest. While these cells could not be used for regeneration of plants, they do offer the convenience of evaluating gene expression in a native cotton cell environment. This study demonstrated that gram quantities of cotton tumor cells expressing the ICP gene could be developed within two months and could be used for expression assays in the laboratory. With a *cry1Ac* Bt gene as an example, insecticidal efficacy was demonstrated in both tobacco (against tobacco hornworm) and cotton (against tobacco budworm) (Figures 3 and 4). It is noteworthy that both tobacco and cotton experiments were completed within three to four months of initiation. Similar efforts to evaluate gene constructs in cotton Steinitz et al. (2002) and other crops (e.g., Williams et al. 1987; Steinitz et al. 1993) were undertaken but they relied on undifferentiated callus cultures, which took more than 16 months to complete. Use of rapidly growing, tumor cells in this study resulted in a considerable saving of time. A similar assay has been reported by Van Wordragen et al. (1993) who used tumor cells of chrysanthemum to evaluate the insecticidal activity conferred by the expression of a *cry1Ab* Bt gene. The lengthy period needed by Steinitz et al. (2002) was partly due to the use of the antibiotic selectable marker, Hygromycin B, which was found to be toxic to Lepidopteran larvae. As a result, they had to subculture the test callus material in the absence of Hygromycin B prior to conducting a meaningful insect-feeding assay. Our own experience with hygromycin selection was similar to their observation (Rajasekaran, unpublished); however, in the present study, no toxic or negative effects of kanamycin (or its analog G418) were observed on the test insect larvae, *Heliothis virescens*.

Tumor cells could be produced in all cotton species and varieties with equal efficiency, thus overcoming the genotype barrier to test the gene constructs (Table 1 and Rajasekaran, unpublished). We are currently using a similar approach to test the antimicrobial effects of different gene constructs (Rajasekaran et al. 1997). Likewise, hairy root cultures initiated following infection with *A. rhizogenes* could be useful in studying root specific gene expression (Hudspeth et al. 1996) or root pathogen (microbial or nematode)-host plant interaction.

In summary, oncogenic strains of Agrobacteria are extremely useful for rapid evaluation of potential gene constructs for input traits. This is especially so in transformation-recalcitrant species, such as cotton, for which the production of transgenic plants usually takes an average of 15 months due to prolonged selection and regeneration procedures.

REFERENCES

An G, Ebert PR, Mitra A and Ha SB (1988) Binary vectors. In: Gelvin SB, Schilperoort RA and Verma DPS (eds.) Plant Molecular Biology Manual. Kluwer Academic Publishers, Dordrecht, pp A3/1-19.

Bedbrook J (1981) A plant nuclear DNA preparation procedure. *PMB Newsletter*, Vol II: 24.

Chilton SW, Temple J, Matzke M and Chilton M-D (1984) Succinamopine: a new crown gall opine. *J. Bacteriol.* 157: 357-362.

Firoozabady E, DeBoer D, Merlo D, Halk E, Amerson L, Rashka K and Murray E (1987) Transformation of cotton (*Gossypium hirsutum* L.) by *Agrobacterium tumefaciens* and regeneration of transgenic plants. *Plant Mol. Biol.* 10: 105-116.

Horsch RB, Fry JE, Hoffmann NL, Rogers SG and Fraley, RT (1985) A simple and general method for transferring genes into plants. *Science* 227: 1229-1231.

Hudspeth RL, Hobbs SL, Anderson DM, Rajasekaran K and Grula JW (1996) Characterization and expression of metallothionein-like genes in cotton. *Plant Mol. Biol.* 31: 701-705.

Murashige T and Skoog F (1962) A revised medium for rapid growth and bioassays with tobacco tissue cultures. *Physiol. Plant.* 15: 473-479.

Rajasekaran K, Cary JW, DeLucca AJ, Jacks TJ, Lax AR, Chen Z, Chlan C, Jaynes J and Cleveland TE (1997) *Agrobacterium*-mediated transformation and analysis of cotton expressing antifungal peptides. In: Robens J (ed.) Proc. Aflatoxin Elimination Workshop, USDA, ARS, p. 44.

Rajasekaran K, Cary JW, Jacks TJ, Stromberg K and Cleveland TE (2000) Inhibition of fungal growth *in planta* and *in vitro* by transgenic tobacco expressing a bacterial nonheme chloroperoxidase gene. *Plant Cell Rep.* 19: 333-338.

Rajasekaran K, Chlan C and Cleveland TE (2001) Tissue culture and genetic transformation of cotton. In: Jenkins JJ and Saha S (eds.) Genetic Improvement of Cotton: Emerging Technologies. Science Publishers, Enfield, NH. pp. 269-290.

Rajasekaran K, Grula JW, Hudspeth RL, Pofelis S and Anderson DM (1996) Herbicide-resistant Acala and Coker cottons transformed with a native gene encoding mutant forms of acetohydroxyacid synthase. *Mol. Breed.* 2: 307-319.

Rothstein SJ, Lahners KN, Lotstein RJ, Carozzi NB, Jayne SM and Rice DA (1987) Promoter cassettes, antibiotic-resistant genes, and vectors for plant transformation. *Gene* 53: 153-161.

Sambrook J, Fritsch EF and Maniatis T (1989) Molecular Cloning–A Laboratory Manual. 2nd ed. Cold Spring Harbor Laboratory Press, New York.

Steinitz B, Gafni Y, Cohen Y, Diaz JP, Tabib Y, Levski S and Navon A (2002) Insecticidal activity of a *Cry*IAc transgene in callus derived from regeneration-recalcitrant cotton (*Gossypium hirsutum* L.). *In Vitro Cell. Dev. Biol.–Plant* 38: 247-251.

Steinitz B, Navon A, Berlinger MJ and Klein M (1993) Expression of insect resistance in vitro-derived callus tissue infested with Lepidopteran larvae. *J. Plant Physiol.* 142: 480-484.

U.S. Department of Agriculture, National Agricultural Statistics Service (USDA-NASS). (2001) Agricultural Chemical Usage–2000 Field Crops Summary. <*http://usda.mannlib. cornell.edu/reports/nassr/other/pcu-bb/agcs0501.pdf*>.

Van Wordragen MF, Honee G and Dons HJM (1993) Insect-resistant chrysanthemum calluses by introduction of *Bacillus thuringiensis* crystal protein gene. *Transgenic Res.* 2: 170-180.

Wilkins TA, Rajasekaran K and Anderson D (2000) Cotton biotechnology. *Crit. Rev. Plant Sci.* 19: 511-550.

Williams WP, Bukely PM and Davis FM (1987) Tissue culture and its use in investigations of insect resistance in maize. *Agric. Ecosyst. Environ.* 18:185-190.

Cauliflower Plants
Expressing a *cry1C* Transgene
Control Larvae of Diamondback Moths
Resistant or Susceptible to Cry1A,
and Cabbage Loopers

Jun Cao
Aigars Brants
Elizabeth D. Earle

SUMMARY. A synthetic *Bacillus thuringiensis* (Bt) *cry1C* gene under the control of the 35S CaMV promoter was introduced into cauliflower (*Brassica oleracea* var. *botrytis*) by *Agrobacterium tumefaciens*-mediated transformation with hygromycin selection. A total of 35 transgenic

Jun Cao (E-mail: jc58@cornell.edu) is Research Associate, Aigars Brants (E-mail: aigars@attbi.com) is Postdoctoral Fellow, and Elizabeth D. Earle (E-mail: ede3@cornell.edu) is Professor, Department of Plant Breeding, Cornell University, Ithaca, NY 14853-1902.

The authors thank Dr. N. Strizhov (Max Planck Institute, Cologne) for providing pNS₆, Dr. A.M. Shelton (New York Agricultural Experiment Station, Geneva, NY) for supplying the diamondback moth eggs, and Dr. P.R. Hughes (Boyce Thompson Institute, Ithaca, NY) for supplying the cabbage looper larvae. The authors are grateful to Kerrie Seberg for her contributions to the assays of the *cry1C* cauliflower plants and to Dr. Ali Alan and Dr. J.-Z. Zhao for critical reading of the manuscript.
This work was supported by USDA Grant 99-35302.

[Haworth co-indexing entry note]: "Cauliflower Plants Expressing a *cry1C* Transgene Control Larvae of Diamondback Moths Resistant or Susceptible to Cry1A, and Cabbage Loopers." Coa, Jun, Aigars Brants, and Elizabeth D. Earle. Co-published simultaneously in *Journal of New Seeds* (Food Products Press, an imprint of The Haworth Press, Inc.) Vol. 5, No. 2/3, 2003, pp. 193-207; and: *Bacillus thuringiensis: A Cornerstone of Modern Agriculture* (ed: Matthew Metz) Food Products Press, an imprint of The Haworth Press, Inc., 2003, pp. 193-207. Single or multiple copies of this article are available for a fee from The Haworth Document Delivery Service [1-800-HAWORTH, 9:00 a.m. - 5:00 p.m. (EST). E-mail address: docdelivery@haworthpress.com].

http://www.haworthpress.com/store/product.asp?sku=J153
10.1300/J153v05n02_06

plants were regenerated from six cultivars (Freemont, Candid Charm, Snow Crown, Cumberland, Majestic, and Cashmere) with average transformation efficiency of 0.3% to 6.4%. All the hygromycin-resistant transformants also carried the Bt gene, as shown by PCR with primers specific to the *cry1C* gene. ELISA analysis showed that the levels of Cry1C protein in independent transformants varied widely, from 0 to 0.2% of total soluble protein. The majority of the plants (61%) produced a high level of Cry1C protein (> 1000 ng mg^{-1} proteins). Insect bioassays demonstrated that plants producing Cry1C protein effectively controlled larvae of diamondback moths (*Plutella xylostella*), including ones resistant to Cry1A protein, as well as larvae of cabbage loopers (*Trichoplusia ni*). These *cry1C* cauliflower plants will be useful for further studies, especially in comparisons with cauliflower plants carrying the same *cry1c* gene under control of a light-inducible promoter. *[Article copies available for a fee from The Haworth Document Delivery Service: 1-800-HAWORTH. E-mail address: <docdelivery@haworthpress.com> Website: <http://www.HaworthPress.com> © 2003 by The Haworth Press, Inc. All rights reserved.]*

KEYWORDS. *Bacillus thuringiensis, Brassica oleracea* var. *botrytis, Plutella xylostella*, resistance management, transformation, transgenic, *Trichoplusia ni*

INTRODUCTION

Cauliflower (*Brassica oleracea* var. *botrytis*) is an economically important vegetable whose annual world production has steadily increased to over 15.7 million tons in 2001, of which 74% was produced in China and India (FAOSTAT 2001). A major problem in cauliflower cultivation is insect pests. The diamondback moth (DBM, *Plutella xylostella*) poses the greatest threat to cauliflower and other crucifer vegetables, sometimes causing more than 90% crop loss (Verkerk and Wright 1996). The annual costs to control DBM have been estimated at over US$1 billion (Talekar and Shelton 1993). DBM are particularly difficult to control because of their high reproductive rate and ability to become resistant to most insecticides, including even the newest classes of insecticides (Talekar and Shelton 1993; Zhao et al. 2002a). Furthermore, use of insecticides is becoming increasingly problematic for other reasons, such as costs, government regulations, health hazards of some insecticides, and consumer concern about pesticide residues.

Advances in development of gene transfer techniques have dramatically increased our ability to introduce agronomically important traits, including insect protection. Transformation of cauliflower has been reported by several laboratories using either *Agrobacterium tumefaciens* (David and Tempe 1988; Passelegue and Kerlan 1996; Bhalla and Smith 1998), *A. rhizogenes* (Christey et al. 1997; Puddephat et al. 2001) or direct DNA transfer (Mukhopadhyay et al. 1991). In recent years, heterologous insecticidal genes have been introduced into cauliflower. Ding et al. (1998) produced transgenic cauliflower plants carrying a trypsin inhibitor gene isolated from sweet potato. Molecular analyses and bioassays demonstrated that the plants producing the trypsin inhibitor inhibited development and increased mortality of larvae of *Spodoptera litura* and DBM. Kuvshinov et al. (2001) reported that introduction of a synthetic *cry9Aa* Bt gene into cauliflower provided complete control of susceptible, Cry1A^R and Cry1C^R DBM larvae.

Proteins encoded by *cry* genes from *B. thuringiensis* are known as the most effective proteinaceous insecticides used in agriculture (Entwistle et al. 1993). Development of Bt-transgenic crops has revolutionized our ability to control some insect pests. However, there are concerns that sustained efficacy of Bt crops will be endangered by the development of resistance in populations of pests that they are intended to control (Roush and Shelton 1997). Benefits of Bt crops will be sustainable only if insect resistance management is properly executed. We have previously produced Bt-transgenic broccoli expressing one or several Bt genes (Metz et al. 1995; Cao et al. 1999, 2002) for use in empirical studies of refuge and pyramiding strategies for resistance management (Shelton et al. 2000; Tang et al. 2001; Zhao et al. unpublished data).

We are now interested in additional approaches to resistance management and increased acceptability of Bt-crucifer crops. As a first step toward that goal, we report here the production of cauliflower plants with constitutive expression of a *cry1C* gene and the effectiveness of these plants in control of several important Lepidopteran pests of crucifers, including DBM that have developed resistance to Cry1A proteins.

MATERIALS AND METHODS

Plant materials: Six cauliflower (*Brassica oleracea* var. *botrytis* L.) cultivars (Candid Charm, Cashmere, Cumberland, Freemont, Majestic, Snow Crown) were used. Seeds were sterilized in 70% ethanol for 5 min, followed by 30% Clorox for 10 min, and rinsed five times with sterile water. Sterilized seeds were placed on MS (Murashige and Skoog 1962) medium containing no growth regulators, 1% sucrose, pH 5.8, solidified with 2.2 g gelrite (Sigma, St. Louis, MO) and maintained in a culture room at 25°C with a 16/8 light/dark photoperiod for germination.

Agrobacterium strain and plasmid: Agrobacterium tumefaciens strain ABI harboring a binary vector pNS_6 (Strizhov et al. 1996; Cao et al. 1999) was used. The binary vector contains a synthetic *cry1C* gene controlled by a CaMV 35S promoter with four enhancer regions and the hygromycin phosphotransferase (*hpt*) gene for convenient selection of hygromycin resistant plants. Prior to use in transformation the *Agrobacteria* were grown overnight at 28°C with vigorous shaking in LB medium supplemented with 200 mg/L hygromycin.

Plant Transformation

Explants consisting of 1 cm hypocotyl sections and cotyledonary petioles (cotyledons plus some of the petiole but excluding the apical meristem) were collected from 8-10 day old germinated seedlings and dipped individually for about 3 seconds into a 1:10 dilution of an overnight *Agrobacterium* culture ($A_{600} = 1.4$) with MS liquid medium. Excess *Agrobacterium* solution was then blotted from the explants on sterile paper towels. Explants (about 20 per plate) were co-cultivated for three days on MS medium containing 3% sucrose, 3.7 mg/L benzyladenine, pH 5.8, solidified with 0.8% Phytagar (Invitrogen, Carlsbad, CA). Hypocotyls were placed horizontally on the medium while cotyledonary petioles were positioned vertically. Plates were sealed with Micropore tape (3M, St. Paul, MN). After co-cultivation, explants were first transferred to plates of the same medium plus 300 mg/L Timentin (SmithKline Beecham Pharmaceuticals, Philadelphia, PA) for 7 days to kill *Agrobacteria* and then to plates of selection medium containing the previous components plus 10 mg/L hygromycin. Explants were subcultured to fresh selection medium every two weeks. Green shoots regenerated on selection medium were transferred to Magenta® boxes of rooting medium (MS medium, 1% sucrose, 300 mg/L Timentin, 10 mg/L hygromycin, pH 5.8). All *in vitro* cultures were maintained in the same light and temperature conditions used for seed germination. Rooted plantlets were transferred to Cornell soil-less mix (Boodley and Sheldrake 1973) in 4.4-cm plastic pots and covered with a plastic bag to maintain a high level of humidity. Holes were gradually cut in the bag over 7-10 days before it was completely removed. Some plants were transferred to larger pots and placed in the greenhouse for further development and recovery of seeds.

DNA isolation and PCR analysis: Total cellular DNA isolation from leaf tissues and polymerase chain reaction (PCR) analysis of genomic DNA were carried out as previously described by Cao et al. (1999). The PCR primers used for the *cry1C* gene were as follows: forward: 5'-GGAGAAAGATGGGGATTG-3'; reverse: 5'-AACTCGTGCATCCCTACT-3'.

Protein isolation and ELISA: Isolation of total soluble proteins from leaf tissues of *in vitro* grown plants and subsequent enzyme-linked immunosorbant assays (ELISA) for detection and quantification of Cry1C protein were carried out using kits from EnviroLogix Inc. (Portland, ME). Proteins from an untransformed cauliflower plant were used as the control. The concentration of total soluble proteins was determined according to the BioRad protein assay (catalog No. 5000-0006). The amount of Cry1C protein present in transgenic plants was calculated from standard curves obtained with Cry1C protein (EnviroLogix) after the O.D. value of the untransformed control was subtracted. The Cry1C protein level was expressed as ng per mg total soluble protein.

Insects: Susceptible and Cry1A-resistant (Cry1A^R) colonies of DBM and cabbage loopers were used for bioassays. The susceptible colony, designated susceptible Geneva 88 (Perez et al. 1997), was collected in 1988 from cabbage plants at the New York State Agricultural Experiment Station, Robbins Farm, Geneva, NY and then maintained on a wheat germ-casein artificial diet (Shelton et al. 1991). The Cry1A^R colony was collected in 1994 from commercial crucifer fields in Loxahatchee, FL and reared on *cry1Ac* broccoli plants (Metz et al. 1995) for 106 generations (Zhao et al. 2002b). Larvae of *Trichoplusia ni* (cabbage looper) were obtained from laboratory colonies maintained as described by Hughes and Wood (1981).

Bioassays: Insect bioassays were performed using detached leaves of *in vitro* grown plants (Cao et al. 1999). Ten DBM larvae freshly hatched from eggs of either susceptible or Cry1A^R DBM were placed on leaves maintained in a Polar plastic cup (Bio Serv, Frenchtown, NJ) containing 1% water agar under a 16/8 (light/dark) regime at 25°C. Two replicates were performed for each bioassay for each type of insect. Untransformed Freemont plants were used as controls. Leaf damage and larval mortality were scored after 4 days. Similar procedures were used in bioassays of detached leaves of *in vitro* grown plants with neonate larvae of *Trichoplusia ni*. For bioassays of intact plants, 10 2nd instar larvae of Cry1A^R DBM were applied to leaves of *in vitro* grown plants with 2 larvae per leaf. Results were scored after 5 days.

RESULTS

Regeneration of Hygromycin-Resistant Plantlets

Within 3 to 4 weeks after explants were placed on selective MS medium containing 10 mg/L hygromycin, green calli emerged from the cut ends of some inoculated explants (but not from uninoculated controls). Shoots were subsequently regenerated from some of the green calli in the following weeks. Roots started to develop 3 to 4 weeks after transfer of the shoots to rooting medium containing 10 mg/L hygromycin.

Two sets of transformation experiments were carried out (Table 1). All cultivars expressed some capacity for regeneration on selective medium, but transformation efficiency varied. Freemont, Candid Charm, and Snow Crown showed capacity to produce *cry1C* plants from both hypocotyl and petiole explants, with Freemont showing the most consistent results. Of the two types of explants tested, hypocotyls of all three cultivars gave higher transformation efficiency than petioles. Majestic, Cumberland, and Cashmere had lower regeneration capacity on selective medium and produced only a few hygromycin-resistant plantlets from petiole explants. Fewer hypocotyl explants than petiole explants were tested, however.

TABLE 1. *Agrobacterium tumefaciens*-Mediated Transformation of 6 Cultivars of Cauliflower with a Synthetic *cry1C* Transgene

Cultivar	Cotyledonary petioles					Hypocotyls				
	Explants		Hygromycin^R plantlets		Transformation efficiency*	Explants		Hygromycin^R plantlets		Transformation efficiency*
	I	II	I	II	(%)	I	II	I	II	(%)
Freemont	180	97	4	5	3.2	140	45	4	3	3.8
Candid Charm	99	95	1	0	0.5	59	48	5	2	6.5
Snow Crown	117	44	3	0	1.9	93	20	1	1	1.8
Majestic	175	60	0	3	1.3	21	32	0	0	0.0
Cumberland	103	60	2	0	1.2	66	20	0	0	0.0
Cashmere	180	80	1	0	0.4	21	37	0	0	0.0

* Aggregate transformation efficiency in the two experiments, I and II.

Molecular Analyses of cry1C Cauliflower Transformants

Genomic PCR

Selected hygromycin-resistant Freemont, Candid Charm, Snow Crown, Cumberland, and Majestic plants were tested by PCR with primers targeting the coding region of the *cry1C* gene. The expected 1.0 kb band was present in PCR products amplified from genomic DNA of all putative transgenic plants while the non-transgenic control was negative (Figure 1). The PCR analysis demonstrated that the hygromycin selection successfully excluded non-transgenic plants.

ELISA

ELISA analysis showed that all transgenic plants tested produced Cry1C protein except for one Freemont transformant, FH-4 (Table 2). The amount of the protein produced varied among the 23 independent transformants. Fourteen (61%) produced high levels of Cry1C protein (> 1000 ng per mg total soluble proteins) with the highest level reaching 2,321 ng Cry1C mg^{-1} proteins or 0.23%. About 13% of the plants showed moderate (728 and 898 ng Cry1C mg^{-1} protein) or moderately low (162 and 497 ng Cry1C mg^{-1} protein) production of Cry1C protein. Two Freemont transformants had a low level of Cry1C production (< 100 ng mg^{-1} protein).

FIGURE 1. PCR analysis of selected hygromycin-resistant cauliflower plants for the *cry1C* gene. Lanes: 1, pNS$_6$ carrying the *cry1C* gene; 2, untransformed control; 3-15, independent *cry1C* cauliflower plants where 3-7 are Freemont, 8-11 are Candid Charm, 12-13 are Snow Crown, 14 Majestic, and 15 Cumberland; 16, 1 kb DNA ladder. Lane 6, DNA sample from F-H4, which showed no detectable Cry1C protein.

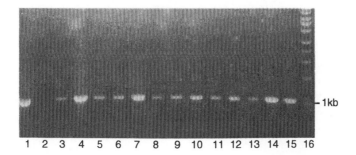

TABLE 2. Production of Cry1C Protein in *cry1C* Cauliflower Plants

Cultivar	Level of Cry1C protein	# of plants	Cry1C protein (ng/mg*)
Freemont	High	4	1098-1739
	Moderate-low	3	162-497
	Low	2	59; 95
	Undetectable	1	0
Candid Charm	High	6	1050-2321
	Moderate	2	728; 898
Snow Crown	High	2	1026; 1342
	Moderate	1	759
Majestic	High	1	2194
Cumberland	High	1	1949

* ng Cry1C protein/mg total soluble protein, evaluated by ELISA.

Insect Bioassays of cry1C Cauliflower Plants

Detached leaf assays were conducted on 11 *cry1C* transformants (4 Freemont, 4 Candid Charm, 1 each of Snow Crown, Majestic and Cumberland) as well as an untransformed Freemont control (Table 3). Neonates of susceptible and Cry1A^R DBM fed voraciously on the control leaves and caused 80% defoliation within 4 days after hatching (Figure 2, top). About 80% of the larvae survived and grew to the 2nd instar stage by day 4. In contrast, susceptible and Cry1A^R neonates stopped feeding and showed 100% mortality within two days after being placed on transgenic leaves. This was true on plants that produced high, moderate, or moderately low levels of Cry1C protein. Leaves from tested plants remained unaffected (Figure 2, bottom). Leaves of plant FR-H2, which produced a low level of Cry1C protein, also caused 100% mortality of susceptible and Cry1A^R DBM larvae, although it suffered slight leaf damage ($< 5\%$), probably from their initial feeding. The one plant (FR-H4) that produced no detectable Cry1C protein suffered severe leaf damage (60-85% defoliation) by both types of DBM and allowed larvae to advance to 2nd instar stage, as on the control.

We also tested the insecticidal activity of the Cry1C-producing plants with *Trichoplusia ni* neonates. Similar results were obtained: transgenic plants producing high, moderate or moderately low levels of Cry1C protein caused rapid and complete mortality of *T. ni* larvae with

TABLE 3. Control of Neonate Larvae by *cry1C* Cauliflower Plants

Cultivar	Plant	Cry1C protein level	Susceptible DBM Defoliation (%)*	Mortality (%)*	Cry1AR DBM Defoliation (%)*	Mortality (%)*	Cabbage looper Defoliation (%)*	Mortality (%)*
Freemont	Control	None	80	10	75-90	25	90-100	0
	FR-H4	None	80-85	15	60-85	15	85-90	5
	FR-H2	Low	3-5	100	0	100	2-15	90
	FR-H5	Mod-low	0	100	0	100	0	100
	FR-P2	High	0	100	0	100	0	100
Candid Charm	CA-P1	Moderate	0	100	0	100	0	100
	CA-H4	High	0	100	0	100	0	100
	CA-H6	High	0	100	0	100	0	100
	CA-H7	High	0	100	0	100	0	100
Snow Crown	SN-P3	High	0	100	0	100	0	100
Majestic	MA-P1	High	0	100	0	100	0	100
Cumberland	CU-P1	High	0	100	0	100	0	100

* Defoliation and insect mortality from detached leaf assays evaluated after 4 days.

no defoliation (Table 3). A few larvae (10%) on plant FR-H2, which had a low level of Cry1C protein, survived to day 4 and caused slight defoliation; however, growth of these larvae was inhibited and they were dead a few days later.

To determine the effects of *cry1C* plants on older DBM larvae, we placed 2nd instar Cry1A^R larvae on two intact *in vitro* grown plants producing high levels of Cry1C protein (FR-P2 and CA-H7) and on control plants. After 5 days, extensive feeding damage was observed on untransformed leaves, actively feeding larvae were visible on the damaged leaves, and all larvae survived and advanced to 3rd or 4th instar stage (Figure 3, left). The transgenic plants showed no leaf damage and caused 100% mortality of the larvae (Figure 3, right).

DISCUSSION

As a first step toward our goal of developing various types of Bt cauliflower for further studies, we established a system for regeneration of transgenic cauliflower plants. The procedure described provides a simple and reproducible method for *A. tumefaciens*-mediated transformation with hygromycin selection. Plants selected for hygromycin resistance all contained the *cry1C* gene as well. The procedure was successful with

FIGURE 2. Bioassay of detached leaves from control and cry1C transgenic cauliflower plants. Ten neonate larvae of susceptible and Cry1A^R DBM were applied to each leaf and allowed to feed for 4 days. Top row: leaves from untransformed Freemont control; bottom row: leaves from transgenic Freemont plant. Left in each row: susceptible larvae applied; right in each row: Cry1A^R larvae applied.

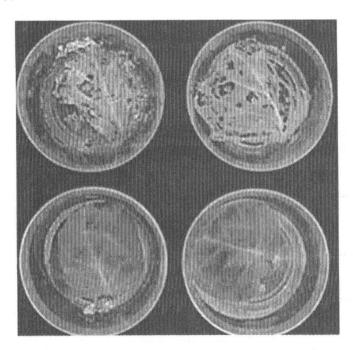

all commercial genotypes tested. The best results were obtained using hypocotyl explants from cultivars Freemont and Candid Charm, with a transformation efficiency as high as 8% (Candid Charm hypocotyl explants, Experiment I). Most *cry1C* plants survived the transfer to soil, and plants grown in the greenhouse showed normal morphology and seed production.

Many laboratories have reported variation in the amounts of Bt proteins produced in independently transformed Bt plants (Stewart et al. 1996; Wunn et al. 1996). We observed a similar wide range of Cry1C protein levels among our independent transformants. The majority produced a high level of the protein (> 0.1% of total soluble proteins) with one plant reaching 0.23%. This was comparable not only to the high levels of Cry1C protein in our *cry1C* broccoli plants (Cao et al. 1999)

FIGURE 3. Bioassay of *in vitro* grown *cry1C* cauliflower plants. Intact plants were assayed for control of Cry1A^R 2nd instar DBM larvae. Larvae were allowed to feed on each plant for 5 days. Left: untransformed Freemont plant. Right: transgenic Freemont plant.

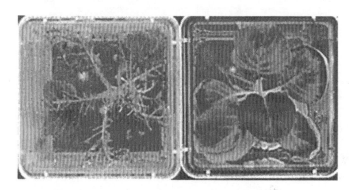

but also to other reports of Bt-transgenic plants. Using the same *cry1C* gene, Strizhov et al. (1996) produced *cry1C* alfalfa and tobacco with the expression of the Cry1C protein up to 0.2% of total soluble protein, which provided protection against Egyptian cotton leaf worm (*Spodoptera littoralis*) and the beet armyworm (*Spodoptera exigua*). Stewart et al. (1996) observed the production of Cry1Ac protein at over 1,300 ng per mg total extractable proteins in transgenic canola that completely controlled the Brassica specialists, DBM and cabbage loopers. All of our cauliflower plants that produced Cry1C protein showed effective control of these two major insect pests of crucifers.

Evolution of resistance to Cry1A in field populations of DBM has been documented in Hawaii (Tabashnik et al. 1990), Florida (Shelton et al. 1993a, 1993b), Central America (Perez and Shelton 1997), and China (Zhao et al. 1993) after heavy use of field sprays of Bt ssp. *kurstaki* formulations. Resistance in these populations of DBM showed little cross-resistance to Cry1C (Liu and Tabashnik 1997; Zhao at al. 2001). Our cauliflower plants with constitutive high expression of Cry1C protein showed complete control of the Cry1A^R DBM. Such cauliflower plants increase our ability to control Lepidopteran pests of this crop, including DBM that have developed resistance to the protein encoded by the widely used *cry1Ac* gene.

Although Bt crops provide benefits in controlling some major agricultural insect pests, proper insect resistance management must be implemented for sustainable use of these crops. Several strategies for limiting the development of resistant insect populations and prolonging the utility of Bt have been proposed. These include high dose of a single gene with a refuge, pyramiding of multiple insect control genes, and regulation of spatial or temporal expression of insect resistant genes. Our previous research on broccoli provided experimental data supporting the current EPA endorsed resistance management strategy (high-dose/single gene with refuge [US EPA 1998]) and also the more advanced insect control strategy of pyramiding of multiple insect resistance genes (Cao et al. 2002; Zhao et al. unpublished). Furthermore, our research has provided data about the potential use of a chemically inducible promoter to regulate Bt expression (Cao et al. 2001).

In light of these developments, we have now produced *cry1C* cauliflower plants with high efficacy in controlling major crucifer insects. These plants can serve as recipients for additional Bt or other insect resistance genes for multi-gene pyramiding. We have recently succeeded in transforming cauliflower from leaf explants (rather than seedling explants) through *Agrobacterium tumefaciens*-mediated transformation. Thus, it should be possible to add *cry1A* genes available in our laboratory to the *cry1C* cauliflower plants by transformation of leaf explants, rather than by sexual crosses as done previously (Cao et al. 2002). Regeneration of multi-*cry1* gene cauliflower plants would not only protect cauliflower from attack by both cry1A^R and cry1C^R DBM (Cao et al. 2002) but would also delay development of resistance in insect populations with low frequencies of alleles to these proteins (Zhao et al. unpublished data).

Use of Bt crops involves many factors other than effective insect control. Regulatory, economic, environmental, health, and public acceptance issues must also be considered. Elimination of transgene expression from edible parts of a plant may increase public acceptance of transgenic crops and safety to non-target organisms. For this reason, we plan to compare cauliflower plants with constitutive expression of the cry1C gene (described in this paper) with more recently obtained cauliflower plants carrying the same *cry1C* gene under control of a light-inducible (ST-LSI) promoter. Such a promoter would be expressed in green leaves but not in the curds or roots. These studies will allow us to determine whether acceptable insect control can be achieved without expression of Bt proteins in the commercial product. In addition, it will be of interest to compare release of Bt protein into the soil from roots of plants with constitutive or light-induced expression of the *cry1C* gene. Concerns have been raised about the presence of Bt proteins in the soil around transgenic plants (Saxena et al. 1999; Stotzky 2000). While there is little evidence that this is a hazard to non-target organisms in the rhizosphere (Saxena and Stotzky 2001), limiting gene expression to aboveground parts may be desirable for protection from defoliating pests and at the same time minimizing any cost of transgene expression.

In conclusion, we have transferred a *cry1C* gene into cauliflower via *Agrobacterium tumefaciens*-mediated transformation. A total of 35 *cry1C* plants were regenerated from six cultivars. Most of them produced over 1,000 ng Cry1C protein per mg total soluble protein and effectively controlled larvae from susceptible and Cry1A^R DBM colonies as well as cabbage loopers.

REFERENCES

Bhalla PL, Smith N (1998) *Agrobacterium tumefaciens*-mediated transformation of cauliflower, *Brassica oleracea* var. *botrytis*. Molecular Breeding 4:531-541.

Boodley JW, Sheldrake R Jr (1973) Cornell peat-like mixes for commercial plant growing. NY Agricultural Experiment Station Agricultural Information Bulletin 43.

Cao J, Tang JD, Strizhov N, Shelton AM, Earle ED (1999) Transgenic broccoli with high levels of *Bacillus thuringiensis* Cry1C protein control diamondback moth larvae resistant to Cry1A or Cry1C. Molecular Breeding 5:131-141.

Cao J, Shelton AM, Earle ED (2001) Gene expression and insect resistance in transgenic broccoli containing a *Bacillus thuringiensis cry1Ab* gene with the chemically inducible PR-1a promoter. Molecular Breeding 8:207-216.

Cao J, Zhao JZ, Tang JD, Shelton AM, Earle ED (2002) Broccoli plants with pyramided *cry1Ac* and *cry1C* Bt genes control diamondback moths resistant to Cry1A and Cry1C proteins. Theoretical and Applied Genetics:105:258-264.

Christey MC, Sinclair BK, Braun RH, Wyke L (1997) Regeneration of transgenic vegetable brassicas (*Brassica oleracea* and *B. campestris*) via Ri-mediated transformation. Plant Cell Reports 16:87-593.

David C, Tempe J (1988) Genetic transformation of cauliflower (*Brassica oleracea* L. var. *botrytis*) by *Agrobacterium rhizogenes*. Plant Cell Reports 7:88-91.

Ding LC, Hu CY, Yeh KW, Wang PJ (1998) Development of insect-resistant transgenic cauliflower plants expressing the trypsin inhibitor gene isolated from local sweet potato. Plant Cell Reports 17:854-860.

Entwistle PF, Cory JS, Bailey MJ, Higgs S (eds) (1993) *Bacillus thuringiensis*, an environmental biopesticide: theory and practice. Wiley, Chichester.

FAOSTAT (2001) http://apps1.faostat.org

Hughes PR, Wood HA (1981) Asynchronous peroral technique for the bioassay of insect viruses. Journal of Invertebrate Pathology 37:154-159.

Kuvshinov V, Koivu K, Kanerva A, Pehu E (2001) Transgenic crop plants expressing synthetic *cry9Aa* gene are protected against insect damage. Plant Science 160: 341-353.

Liu YB, Tabashnik BE (1997) Inheritance of resistance to *Bacillus thuringiensis* toxin Cry1C in the diamondback moth. Applied & Environmental Microbiology 63: 2218-2223.

Metz TD, Roush RT, Tang JD, Shelton AM, Earle ED (1995) Transgenic broccoli expressing a *Bacillus thuringiensis* insecticidal crystal protein: implications for pest resistance management strategies. Molecular Breeding 1:309-317.

Mukhopadhyay A, Töpfer R, Pradham AK, Sodhi, YS, Steinbiss HH, Schell J, Pental D (1991) Efficient regeneration of *Brassica oleracea* hypocotyl protoplasts and high frequency genetic transformation by direct DNA uptake. Plant Cell Reports 10:375-379.

Murashige T, Skoog F (1962) A revised medium for rapid growth and bioassay with tobacco tissue culture. Physiologia Plantarum 15:473-497.

Passelegue E, Kerlan C (1996) Transformation of cauliflower (*Brassica oleracea* var. *botrytis*) by transfer of cauliflower mosaic virus genes through combined cocultivation with virulent and avirulent strains of *Agrobacterium*. Plant Science 113:79-89.

Perez CJ, Shelton AM (1997) Resistance of *Plutella xylostella* (Lepidoptera: Plutellidae) to *Bacillus thuringiensis* Berliner in Central America. Journal of Economic Entomology 90:87-93.

Perez CJ, Tang JD, Shelton AM (1997) Comparison of leaf-dip and diet bioassays for monitoring *Bacillus thuringiensis* resistance in field populations of diamondback moth (Lepidoptera: Plutellidae). Journal of Economic Entomology 90:94-101.

Puddephat IJ, Fenning TM, Barbara DJ, Morton A, Pink DAC (2001) Recovery of phenotypically normal transgenic plants of *Brassica oleracea* upon *Agrobacterium rhizogenes*-mediated co-transformation and selection of transformed hairy roots by GUS assay. Molecular Breeding 7:229-242.

Roush RT, Shelton AM (1997) Assessing the odds: the emergence of resistance to Bt transgenic plants. Nature Biotechnology 15:816-817.

Saxena D, Flores S, Stotzky G (1999) Insecticidal toxin in root exudates from Bt corn. Nature 402:480.

Saxena D, Stotzky G (2001) *Bacillus thuringiensis* (Bt) toxin released from root exudates and biomass of Bt corn has no apparent effect on earthworms, nematodes, protozoa, bacteria, and fungi in soil. Soil Biology & Biochemistry 33:1225-1230.

Shelton AM, Cooley RJ, Kroening MK, Wilsey WT, Eigenbrode SD (1991) Comparative analysis of two rearing procedures for diamondback moth (Lepidoptera: Plutellidae). Journal of Entomological Science 26:17-26.

Shelton AM, Robertson JL, Tang JD, Perez C, Eigenbrode SD, Preisler HK, Wilsey WT, Cooley RJ (1993a) Resistance of diamondback moth (Lepidoptera: Plutellidae) to *Bacillus thuringiensis* subspecies in the field. Journal of Economic Entomology 86:697-705.

Shelton AM, Wyman JA, Cushing NL, Apfelbeck K, Dennehy TJ, Mahr SER, Eigenbrode SD (1993b) Insecticide resistance of diamondback moth (Lepidoptera: Plutellidae) in North America. Journal of Economic Entomology 86:11-19.

Shelton AM, Tang JD, Roush RT, Metz TD, Earle ED (2000). Field tests on managing resistance to Bt-engineered plants. Nature Biotechnology 18: 339-342.

Stewart CNJ, Adang MJ, All JN, Raymer PL, Ramachandran S, Parrott WA (1996) Insect control and dosage effects in transgenic canola containing a synthetic *Bacillus thuringiensis cryIAc* gene. Plant Physiology 112:115-120.

Stotzky G (2000) Persistence and biological activity in soil of insecticidal proteins from *Bacillus thuringiensis* and in bacterial DNA bound on clays and humic acids. Journal of Environmental Quality 29:691-705.

Strizhov N, Keller M, Mathur J, Koncz-Kálmán Z, Bosch D, Prudovsky E, Schell J, Sneh B, Koncz C, Zilberstein A (1996) A synthetic *cry1C* gene, encoding a *Bacillus thuringiensis*-endotoxin, confers *Spodoptera* resistance in alfalfa and tobacco. Proceedings of the National Academy of Sciences U.S.A. 93:15012-15017.

Tabashnik BE, Cushing NL, Finson N, Johnson MW (1990) Field development of resistance to *Bacillus thuringiensis* in diamondback moth (Lepidoptera: Plutellidae). Journal of Economic Entomology 83:1671-1676.

Talekar, NS, Shelton, AM (1993) Biology, ecology and management of the diamondback moth. Annual Review of Entomology 38:275-301.

Tang, JD, Collins HL, Metz TD, Earle ED, Zhao J, Roush RT, Shelton AM. 2001. Greenhouse tests on resistance management of Bt transgenic plants using refuge strategies. Journal of Economic Entomology 94: 240-247.

US EPA (1998) Final Report of the Subpanel on *Bacillus thuringiensis* (Bt) Plant-Pesticides and Resistance Management, Meeting held on February 1-10, 1998.

Verkerk, RHJ, Wright, DJ (1996) Multitrophic interactions and management of the diamondback moth: a review. Bulletin of Entomological Research 86:205-216.

Wunn J, Kloti A, Burkhardt PK, Ghosh Biswas GC, Launis K, Iglesias VA, Potrykus I (1996) Transgenic indica rice breeding line IR58 expressing a synthetic *cry1Ab* gene from *Bacillus thuringiensis* provides effective insect pest control. Bio/Technology 14:171-176

Zhao JZ, Zhu G, Zhu ZL, Wang WZ (1993) Resistance of diamondback moth to *Bacillus thuringiensis* in China. Resistant Pest Management 5:11-12.

Zhao JZ, Li YX, Collins HL, Cao J, Earle ED, Shelton AM (2001) Different cross-resistance patterns in the diamondback moth resistant to *Bacillus thuringiensis* toxin Cry1C. J. Econ. Entomol. 94:1547-1552.

Zhao JZ, Li Y, Collins HL, Gusukuma-Minuto L, Mau RFL, Thompson GD, Shelton AM (2002a) Monitoring and characterization of diamondback moth resistance to spinosad. Journal of Economic Entomology 95:430-436.

Zhao JZ, Li YX, Collins HL, Shelton AM (2002b) Examination of the F_2 screen for rare resistance alleles to *Bacillus thuringiensis* toxins in the diamondback moth (Lepidoptera: Plutellidae). Journal of Economic Entomology 95:14-21.

Expression of the Cry1Ab Protein in Genetically Modified Sugarcane for the Control of *Diatraea saccharalis* (Lepidoptera: Crambidae)

Daniella P. V. Braga
Enrico D. B. Arrigoni
Marcio C. Silva-Filho
Eugênio C. Ulian

Daniella P. V. Braga was affiliated with Seção de Biologia Molecular, Centro de Tecnologia Copersucar, Caixa Postal 162, Piracicaba, SP, Brazil 13400-970 and with the Departamento de Genética, Escola Superior de Agricultura "Luiz de Queiroz," Universidade de São Paulo, Caixa Postal 83, Piracicaba, SP, Brazil 13400-970. She is now affiliated with the Regulamentation Division, Monsanto do Brasil Ltda., São Paulo, SP, Brazil 04578-000.

Enrico D. B. Arrigoni is affiliated with Seção de Entomologia, Centro de Tecnologia Copersucar, Caixa Postal 162, Piracicaba, SP, Brazil 13400-970.

Marcio C. Silva-Filho is affiliated with the Departamento de Genética, Escola Superior de Agricultura "Luiz de Queiroz," Universidade de São Paulo, Caixa Postal 83, Piracicaba, SP, Brazil 13400-970.

Eugênio C. Ulian is affiliated with Seção de Biologia Molecular, Centro de Tecnologia Copersucar, Caixa Postal 162, Piracicaba, SP, Brazil 13400-970.

Address correspondence to: Marcio C. Silva-Filho (E-mail: mdcsilva@carpa.ciagri.usp.br).

The authors thank Syngenta for providing plasmids pCIB4421 and pCIB4426, antibodies and Cry1Ab purified protein; Dr. Henrik Albert (Hawaiian Agricultural Research Center) for providing plasmid pHA9; Dr. Jorge A.G. da Silva and Dr. Maria C. Falco for assistance in experiments and helpful comments; and Daniela C. Volpato and Rosemeire Zem for outstanding technical assistance.

[Haworth co-indexing entry note]: "Expression of the Cry1Ab Protein in Genetically Modified Sugarcane for the Control of *Diatraea saccharalis* (Lepidoptera: Crambidae)." Braga, Daniella P. V. et al. Co-published simultaneously in *Journal of New Seeds* (Food Products Press, an imprint of The Haworth Press, Inc.) Vol. 5, No. 2/3, 2003, pp. 209-221; and: *Bacillus thuringiensis: A Cornerstone of Modern Agriculture* (ed: Matthew Metz) Food Products Press, an imprint of The Haworth Press, Inc., 2003, pp. 209-221. Single or multiple copies of this article are available for a fee from The Haworth Document Delivery Service [1-800-HAWORTH, 9:00 a.m. - 5:00 p.m. (EST). E-mail address: docdelivery@haworthpress.com].

http://www.haworthpress.com/store/product.asp?sku=J153
© 2003 by Taylor & Francis.
10.1300/J153v05n02_07

SUMMARY. Previously, we genetically modified two Brazilian commercial sugarcane (*Saccharum* sp.) varieties (SP80-3280 and SP80-1842) for resistance to the sugarcane borer *Diatraea saccharalis* by co-bombardment transformation with three plasmids: pCIB4421 containing the *Bacillus thuringiensis* gene *cryIAb* regulated by a maize phosphoenolpyruvate carboxylase promoter, pCIB4426 carrying *cryIAb* controlled by a pith promoter, and pHA9 with the *neo* antibiotic resistance gene. The genetically modified plants were submitted to molecular characterization and insect feeding bioassays. Plants also were screened in the field for insect infestation and had their phenotypic characteristics and quantitative traits evaluated. This study presents evidence that the *B. thuringiensis cryIAb* gene can be efficiently expressed in sugarcane plants over an extended growing period, and can confer resistance to the sugarcane borer under field conditions, while not measurably altering numerous material traits of the crop. *[Article copies available for a fee from The Haworth Document Delivery Service: 1-800-HAWORTH. E-mail address: <docdelivery@haworthpress.com> Website: <http://www.HaworthPress.com>* © 2003 by The Haworth Press, Inc. All rights reserved.]

KEYWORDS. *Saccharum* sp., *Bacillus thuringiensis*, genetic transformation, sugarcane borer, *cryIAb*

INTRODUCTION

The sugarcane borer *Diatraea saccharalis* (Fabricius) is a widely distributed pest of sugarcane (*Saccharum* sp.) in Brazil which inflicts severe productivity losses (Falco et al., 2001). In sugarcane plants up to 4 months old, borer larvae cause apical bud death ('dead heart') leading to death of the plant, while in older plants damage leads to lateral bud development, aerial rooting, weight loss, stalk breakage, and entry of opportunistic fungi, causing plant death and heavy production losses (Mendonça, 1996; Parra, 1993).

One strategy to produce insect resistance in sugarcane plants is the introduction of genes that code for products that interfere with insect development and mortality, e.g., the soybean proteinase inhibitors (Pompermayer et al., 2001) and the Cry proteins from *Bacillus thuringiensis* (*Bt*). The Bt proteins are toxic to insects when activated by proteinases and the alkaline pH of the insect midgut. Specific binding to receptors facilitates pore formation in the gut membrane, causing death of the insect in a few hours by disruption of osmotic balance (Brunke and Meeusen, 1991;

Frutos et al., 1999; Gill et al., 1992). Genetically modified sugarcane with a *cry1Ab* gene from *Bt* under control of the CaMV 35S promoter were evaluated in field trials and showed significant larvicidal activity to sugarcane borer (Arencibia et al., 1997). These plants, however, showed a low expression level of Cry1Ab insecticidal crystal protein (ICP). Plants with higher expression levels seemed desirable for field deployment of the trait, with a better chance of long-term protection from infestation.

Previously we transformed Brazilian sugarcane varieties with two different constructs carrying the *Bt cry1Ab* gene encoding for the Cry1Ab protein toxic to Lepidoptera. In order to improve the expression levels of the ICP in vulnerable plant tissues, two promoter elements were used simultaneously in independent constructs: a maize phosphoenolpyruvate carboxylase (PEPC) and a pith promoter. The different promoters were chosen to allow the expression of the ICP in green and non-green aerial parts of the plant, those tissues where the insect typically causes damage. We used microprojectile bombardment to transform two commercial sugarcane cultivars (SP80-1842 and SP80-3280).

In this study we employ Southern Blot analysis to examine the molecular integration of the transgenes into the sugarcane genome. An enzyme-linked immunosorbent assay (ELISA) was used to quantify the level of expression of *Bt* gene in different tissues. Further bioassays and field trials provided evidence that, under experimental conditions, the Cry1Ab protein controls *D. saccharalis* in the field.

MATERIALS AND METHODS

Enzyme-Linked Immunosorbent Assays (ELISA)

ELISAs were performed to detect and quantify the expression of the Cry1Ab protein in transformed plants. Polyclonal antibodies specific for *Bacillus thuringiensis, kurstaki* HD-1 insecticidal proteins, and pure Cry1Ab protein were provided by Dr. Nadine Carozzi (Syngenta). The ELISA protocol used was slightly modified from Koziel et al. (1993) and total protein was quantified by the Bradford (1976) method with bovine albumin serum as standard.

Southern Blot Analysis

The protocols for DNA extraction and digestion, agarose gel electrophoresis, DNA transfer to nylon membranes and radioactive hybridization were the same as used by Da Silva et al. (1993), using RediPrime®️ (Amersham Pharmacia Biotech) as the labeling system. DNA from control and transgenic plants was separately digested with *Hind*III and *Eco*RI to determine the *neo* and *cry1Ab* copy number. The probes were generated by PCR amplification of sequences inside both genes and 25 ng of each DNA was labeled with ^{32}P-dCTP Redivue™️ (APBiotech).

Field Tests

Transgenic and non-transgenic control plants of cultivars SP80-1842 and SP80-3280 were grown on 0.17 ha at Copersucar Experimental Station (Piracicaba, SP, Brazil). Twelve plants from each bombardment event, clonally propagated from single-eyed sets, were planted in 6 m rows with 0.5 m between plants and 1.4 m between rows. After 3 months, each plant was infested with 5 neonate leaf borer larvae and infestation was repeated bi-weekly over a 7 month period. Insect damage was evaluated on 22-month-old plants by harvesting 10 stalks from each 6 m plot. The stalks were weighed and opened longitudinally to count total and damaged internodes. Infestation intensity (I.I.) was calculated as:

I.I. = number of damaged internodes $\times$ 100/total number of internodes

Measurements of brix, fiber, pol, purity, and reducing sugars were done using the same stalks used for infestation evaluation, following the protocols used in the Copersucar breeding trials (Consecana, 2000).

RESULTS AND DISCUSSION

Southern Blotting

To assess the gene copy number in each transformation event, genomic DNA was digested with specific endonucleases, without any restriction site inside the *neo* and *cry1Ab* genes. Each transgene insertion event would then be associated with a certain size fragment of genomic DNA and could be identified by specific probes. This evaluation was effective in verifying genetic integration of the different transgenes, with a very different banding pattern for each. This result was indicative of independent integration of the different transgenes, and was expected since co-transformation was done using different plasmids at the same time. Plants that were first selected on geneticin containing medium and later by kanamycin selection in the greenhouse have the *neo* gene (Figure 1), which was used only as a selection marker with no agronomic importance.

Southern blotting was used for the molecular characterization of 14 plants and was able to distinguish 12 independent transformation events (Figure 1), six events being recognized for each variety with transgene copy numbers ranging from 1 to 12. No correlation was found between these results and protein expression. Transformation through microparticle bombardment is recognized as a method that produces complex integration patterns. Multiple transgene copies, DNA rearrangements, gene silencing and random integration can lead to wide variation in pro-

FIGURE 1. Southern Blot analysis using specific probes for (A) the *neo* gene and (B) the *cry1Ab* gene from *Bt* to evaluate the molecular integration of the inserted genes. It also provides the copy number for each transformation event. Note that some plants are result of the same transformation event (106 and 107; 241 and 242) as inferred from the identical transgene insertion patterns (Ctrl+: positive control; Ctrl−: negative control).

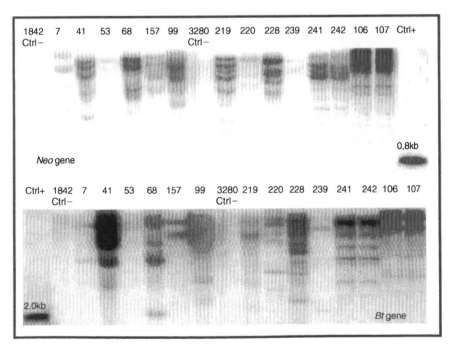

tein expression. For plants 106 and 107, and plants 241 and 242, these pairs appear to be the result of the same transformation events. This is not surprising in experiments where one transformed cell results in large callus clumps from which more than one plant can regenerate.

As for plant 7, which did not present an amplification band when primers specific to the pith promoter were used (Braga et al., 2001), it could be inferred that the *cry1Ab* copies were integrated from the pCIB4421 (PEPC promoter) plasmid. On the other hand, all events showed the presence of the *cry1Ab* and *neo* genes, meaning that co-transformation was very efficient and can be used to introduce other genes into sugarcane. Co-transformation has been successfully reported in sugarcane using the particle bombardment procedure (Birch et al., 1996; Falco et al., 2000). In general, co-transformation employs a selectable marker gene and an agronomically important gene (e.g., the *cry1Ab* gene). Thus, this strategy has two major advantages: (1) it allows the selection of transformed callus and plantlets *in vitro*, and (2) it provides the possibility of breeding out the selectable marker gene. Once selection of transformed plants has been carried out, the marker genes can be eliminated from the plant genome if they are not linked to the other trait of interest and segregate independently (for a review, see Hare and Chua, 2002).

ELISA Test

Twenty-two-month-old plants from field experiments were evaluated for ICP expression using ELISA and, as expected, the Cry1Ab protein was not detected in control plants. Table 1 shows the average of four replications for each plant. These plants had been evaluated in a previous test when they were 10 months old (Braga et al., 2001) when the amount of Cry1Ab ICP measured in the leaves of transgenic plants was found to be much higher than in stalks, leaf sheaths and rind. The only event that had higher expression in leaves at 22 months of age was plant 107, if comparing with our prior sampling of 10-month-old plants. All other plants had lower ICP expression in all sampled tissues at 22 months post-planting in comparison with the first evaluation (at 10 months post-planting). Even though a lower leaf expression of the Cry1Ab protein was detected for the other plants at 22 months they displayed high resistance to borer infestation as indicated by the I.I. for each event. Plant 7 had the lowest expression level in non-green tissues and PCR analysis confirmed that this plant did not have the pith-*cry1Ab*

TABLE 1. Expression of Cry1Ab Protein (ng/mg Total Protein) in Different Tissues and Infestation Intensity of *D. saccharalis* in Different Transgenic and Base Lines at 22 Months Post-Planting

Plant (controls and events)	Tissue Type				I.I. %
	Leaf	Stalk	Rind	Leaf sheath	
SP80-1842 control	0	0	0	0	8.4
7	76.18 ± 6.03	7.36 ± 8.53	20.51 ± 9.1	68.55 ± 38.9	0
41	104.73 ± 68.54	410.84 ± 348.43	132.68 ± 29.69	146.67 ± 65.6	0
53	385.66 ± 245.57	0	0	0	0
99	182.56 ± 52.17	184.55 ± 176.09	147.08 ± 61.31	184.18 ± 20.25	0
157	124.54 ± 44.44	43.83 ± 25.32	50.1 ± 21.03	141.44 ± 37.82	0
SP80-3280 control	0	0	0	0	12.7
107	151.08 ± 213.66	110.06 ± 97.01	157.56 ± 60.46	299.72 ± 102.09	0
219	219.01 ± 53.15	0	0	0	3.1
220	277.11 ± 112.55	142.4 ± 16.77	123.74 ± 21.68	316.3 ± 41.07	0
228	113.27 ± 45.57	0	0	0	7.8
239	315.05 ± 43.61	982.5 ± 866.2	122.71 ± 98.52	368.57 ± 128.28	0

sequence in its genome, while it also had the lowest ICP level in leaf tissue driven only by the PEPC promoter (Table 1).

Different expression levels in specific tissues were expected as a result of promoter efficiency and the effect of the genomic location of transgene insertion, but not because of gene copy number. Leaf tissue was the only tissue showing expression in all the plants analyzed, but by comparing the results in Table 1 and those of Braga et al. (2001) it appears that there was a decrease in leaf expression at 22 months compared with the leaf expression seen at 10 months by Braga et al. (2001). This decline may be attributable to a correlation between PEPC promoter efficiency and tissue age. There is evidence that both promoters drove the expression of the *Bt* gene in stalk tissue, but our data does not allow a firm conclusion as to which is more efficient because both constructs were used in co-transformation.

Plant 239 presented higher expression in leaf sheath and stalk in the 22 months evaluation compared with the evaluation at 10 months which showed PCR amplification for both promoters (Braga et al., 2001) and one copy of the *Bt* gene (Figure 1), with substantial ICP expression in all the tissues analyzed. This shows that molecular characterization is not always predictive of protein expression levels. The variation in protein expression between plants, tissues and replicates (Table 1) shows that the PEPC and pith promoter driven expression is highly variable, providing no correlation with copy number.

It appears that expression of the Bt transgene exclusively in leaf tissue is insufficient to provide complete protection from infestation. Two variety SP80-3280 plants (219 and 228) exhibited no expression in tissues other than leaves, and also had a percentage of infestation intensity greater than zero (even though lower than the control). Incomplete protection from leaf exclusive expression could be due to a number of possible factors including variety morphology, natural insect susceptibility and promoter efficiency. Complete protection appears to require expression in stalk and rind, tissues where sugarcane borer damage is normally found to be concentrated.

The Cry1Ab protein gives excellent control of sugarcane borer under laboratory conditions, similar to other experimental crop/pest systems. Previous observations have shown that transformed rice expressing the *cry1Ab* gene regulated by the maize PEPC and pith promoters interfered with insect development (Datta et al., 1998). These authors showed that approximately 10% of the transgenic plants resulted in 100% mortality of yellow stem borer (*Scirpophaga incertulas*) larvae and the majority produced 70-90% mortality as compared with 10-40% mortality in controls. This level of control was found even in plants showing low expression of the pith promoter construct and also in plants with high expression of the PEPC promoter construct. In Bt transgenic maize, Koziel et al. (1993) performed bioassays with the European corn borer (*Ostrinia nubilalis*) and found high mortality after 48 hours, with some maize plants producing 100% mortality. These authors also found a good correlation between mortality level and the expression of the Cry1Ab protein as detected by ELISA.

Three out of eight SP80-3280 transgenic plants showed some damage, but lower than in control plants. Plants 228 and 219 had expression of the Bt ICP in leaves at lower levels than in most of the transgenic plants analyzed, and no detectable expression in other tissues (Table 1). It appears that in these two plants the observed levels of infestation are a result of the fact that some larvae did not feed on leaves and were able to survive to reach other tissues with insufficient ICP expression. The fact that plants of the same variety producing ICP in all tissues (plants 220, 239, 241 and 242) showed no damage reinforces this interpretation.

Plants 106 and 107 were found to represent the same transformation event (Figure 1) but, due to different expression levels in the same tissue, we have reason to believe that these plants represent mosaicism.

High expression of the *cry1Ab* gene in leaves indicates that, as expected, the PEPC promoter was very efficient in this green tissue. The PEPC promoter should also drive some expression in non-green tissues (probably complemented by expression from the gene driven by the pith promoter) leading to better protection in stalks, rind and leaf sheaths. High expression of a similar synthetic construct using the *cry1Ab* gene with the PEPC, pith and pollen promoters was shown in maize leaves by Koziel et al. (1993). This study showed low levels of ICP expression (probably from the pith promoter) in pith tissue, although the levels were enough to produce insecticidal activity against European corn borer and prevent larval penetration of corn. These authors could not attribute expression levels and patterns to any specific promoter.

Datta et al. (1998) reported a comparison between different constructs used for rice transformation, where the PEPC promoter was the best for Cry1Ab protein production. These authors found that in stem pith tissue, the pith promoter was responsible for low production of the Cry1Ab protein. Consistent with this, our work suggests that the activity of this promoter was much lower in non-green tissues when compared to leaves.

Evaluation of Agronomic Features

It is desirable for genetically modified plants to have the same qualities as conventional plants, but the review by Tu et al. (2000) includes an interesting study with genetically modified rice in which the yield was lower than expected. However, this review also presents recent studies where Bt rice plants were highly efficient in controlling insects without any reduction in yield.

To evaluate the quality of sugarcane produced by these transgenic plants, we measured brix, fiber, pol, purity and reducing sugars (Figure 2) using the same stalks used for the evaluation of infestation of 10-month-old plants. These results show that there were no major differences between transgenic plants and non-transgenic control plants. A reduction in stalk diameter coupled with an increase in the number of stalks for SP80-1842 transgenic plants and a decrease in the number of stalks in SP80-3280 transgenic plants resulted in a slight decrease in

FIGURE 2. Agronomic traits evaluated for the transgenic plants and non-transgenic controls 10 months after planting.

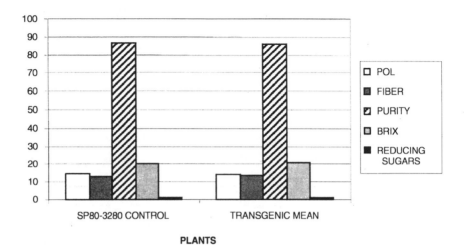

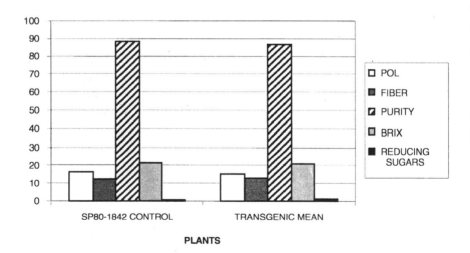

overall stalk weight when compared to the conventional varieties (Table 2). Such results are sometimes observed after tissue culture and probably not due to genetic modification. This phenomenon is also observed in conventional sugarcane plants that are micro-propagated by tissue culture and should disappear after 2 or 3 harvesting cycles.

TABLE 2. Results from the Morphological Assessments at 10 Months After Planting. Diameter Scores Range from 1 (+) to 9 (−)

Plant (controls and events)	Length (meters)	Diameter score	Stalk number	Weight of 10 stalks (Kg)
SP80-1842 control	3.3	4	73	18.9
7	3.35	4	87	17.5
41	3.25	5	80	12.5
53	3.2	6	127	12
99	2.9	6	87	9.5
157	3.3	5	75	16.6
SP80-3280 control	2.9	3	84	17.2
107	3.1	6	57	11.5
219	2.5	5	100	9.5
220	2	5	69	8.5
228	2.6	5	80	14.5
239	2.9	6	60	10.7

The data for brix, fiber, pol, purity and reducing sugars (Figure 2) suggest that there were no detrimental effects that could be associated with transgene insertions. Further trials to evaluate these agronomic traits have been set up and should be able to confirm that the insertion of the *cry1Ab* and *neo* genes have caused no major disruptions to any important agronomic characteristic.

CONCLUSIONS

The results presented in this study confirm the efficacy of the *cry1Ab* gene from *Bacillus thuringiensis* under laboratory and field conditions for sugarcane borer control. All the transgenic plants evaluated showed that the expression of the Cry1Ab protein in different tissues provided highly efficient protection. All transgenic plants showed reduced, and often zero borer damage when compared to conventional varieties under laboratory or field conditions. We achieved high levels of expression of the Bt ICP in two different sugarcane varieties. The material characteristics of the crop appear to be equivalent between transgenic and non-transgenic lines.

REFERENCES

Arencibia, A., R.I. Vazquez, D. Prieto, P. Tellez, E.R. Carmona, A. Coego, L. Hernandez, G.A. DelaRiva and G. SelmanHousein. (1997) Transgenic sugarcane plants resistant to stem borer attack. *Molecular Breeding* 3: 247-255.

Birch, R.G., R. Bower, A. Elliot, B. Potier, T. Franks and G. Cordeiro. (1996) Expression of foreign genes in sugarcane. *Congress of the ISSCT, XXII, 1995 Proceedings* 2: 368-373.

Bradford, M.M. (1976) A rapid and sensitive method for the quantitation of microgram quantities of protein utilizing the principle of protein-dye binding. *Analytical Biochemistry* 72: 248-254.

Braga, D.P.V., E.D.B. Arrigoni, W.L. Burnquist, M.C. Silva-Filho and E.C. Ulian. (2001) A new approach for control of *Diatraea saccharalis* (Lepidoptera: Crambidae) through the expression of an insecticidal Cry1Ab protein in transgenic sugarcane. *Proceedings of the International Society for Sugar Cane Technologists* 24: 331-336.

Brunke, K.J. and R.L. Meeusen. (1991) Insect control with genetically engineered crops. *Trends in Biotechnology* 9: 197-200.

Consecana-SP. (2000) Manual de Instruções. São Paulo, Brasil. 92 pp.

Da Silva, J.A.G., M.E. Sorrels, W.L. Burnquist and S.D. Tanksley. (1993) RFLP linkage map and genome analysis of *Saccharum spontaneum. Genome* 36: 782-791.

Datta, K., A. Vasquez, J. Tu, L. Torrizo, M.F. Alam, N. Oliva, E. Abrigo, G.S. Khush and S.K. Datta. (1998) Constitutive and tissue-specific differential expression of the *Cry1Ab* gene in transgenic rice conferring resistance to rice insect pests. *Theoretical and Applied Genetics* 97: 20-30.

Falco, M.C., A. Tulmann-Neto and E.C. Ulian. (2000) Transformation and expression of a gene for herbicide resistance in a Brazilian sugarcane. *Plant Cell Reports* 19: 1188-1194.

Falco, M.C., P.A.S. Marbach, P. Pompermayer, F.C.C. Lopes and M.C. Silva-Filho. (2001) Mechanisms of sugarcane response to herbivory. *Genetics and Molecular Biology* 24: 113-122.

Frutos, R., C. Rang and M. Royer. (1999) Managing insect resistance to plants producing *Bacillus thuringiensis* toxins. *Critical Reviews in Biotechnology* 19: 227-276.

Gill, S.S., E.A. Cowles and P.V. Pietrantonio (1992) The mode of action of *Bacillus thuringiensis* endotoxins. *Annual Reviews of Entomology* 37: 615-636.

Hare, P.D. and N.H. Chua. (2002) Excision of selectable marker genes from transgenic plants. *Nature Biotechnology* 20: 575-580.

Koziel, M.G., G.L. Beland, C. Bowman, N.B. Carozzi, R. Crenshaw, L. Crossland, J. Dawson, N. Desai, M. Hill, S. Kadwell, K. Launis, K. Lewis, D. Maddox, K. McPherson, M.R. Meghji, E. Merlin, R. Rhodes, G.W. Warren, M. Wright and S.V. Evola. (1993) Field performance of elite transgenic maize plants expressing an insecticidal protein derived from *Bacillus thuringiensis. Biotechnology* 11: 194-200.

Mendonça, A.F., J.A. Moreno, S.H. Risco and I.C.B. Rocha. (1996) As brocas da cana-de-açúcar. In: Mendonça, A. F. (ed.) Pragas da cana-de-açúcar. Insetos e Cia, Maceió. pp. 51-82.

Parra, J.R.P. (1993) Controle das principais pragas da cana-de-açúcar. In: Câmara, G.M.S., Oliveira, E.A.M. (eds.). Produção de Cana de Açúcar. FEALQ, Piracicaba. pp. 184-197.

Pompermayer, P., A.R. Lopes, W.R. Terra, J.R.P. Parra, M.C. Falco and M.C. Silva-Filho. (2001) Effects of soybean proteinase inhibitor on development, survival and reproductive potential of the sugarcane borer, *Diatraea saccharalis*. *Entomologia Experimentalis et Applicata* 99: 79-85.

Tu, J., G. Zhang, K. Datta, C. Xu, Y. He, Q. Zhang, G.S. Khush and S.K. Datta. (2000) Field performance of transgenic elite commercial hybrid rice expressing *Bacillus thuringiensis* delta-endotoxin. *Nature Biotechnology* 18: 1101-1104.

Two Years of Insect Protected
Bt Transgenic Cotton in Argentina:
Regional Field Level Analysis
of Financial Returns and Insecticide Use

Mirta Graciela Elena de Bianconi

SUMMARY. A comparative analysis of conventional and insect protected or "Bt" cotton varieties in farmers fields in the Argentine cotton production area during the 1999/00 and 2000/01 growing seasons was conducted. Use of Bt cotton provided farmers with benefits such as increased crop yields and a considerable reduction in the number of insecticide applications. The average additional benefit obtained in both growing seasons was 57.98 $/ha, equivalent to 16 US$ (July 2002). *[Article copies available for a fee from The Haworth Document Delivery Service: 1-800-HAWORTH. E-mail address: <docdelivery@haworthpress.com> Website: <http://www.HaworthPress.com> © 2003 by The Haworth Press, Inc. All rights reserved.]*

KEYWORDS. Transgenic Bt cotton, economic benefit, direct cost incomes, insecticide use, Argentina

Mirta Graciela Elena de Bianconi is Agronomist, Research Area for Ecology and Management, Economy Section, Agricultural Experimental Station, National Institute of Agricultural Technology (INTA), Saenz Peña, Chaco, Argentina (E-mail: gelena@chaco.inta.gov.ar).

Financial support and valuable information from INTA and Monsanto Company is gratefully acknowledged.

[Haworth co-indexing entry note]: "Two Years of Insect Protected Bt Transgenic Cotton in Argentina: Regional Field Level Analysis of Financial Returns and Insecticide Use." de Bianconi, Mirta Graciela Elena. Co-published simultaneously in *Journal of New Seeds* (Food Products Press, an imprint of The Haworth Press, Inc.) Vol. 5, No. 2/3, 2003, pp. 223-235; and: *Bacillus thuringiensis: A Cornerstone of Modern Agriculture* (ed: Matthew Metz) Food Products Press, an imprint of The Haworth Press, Inc., 2003, pp. 223-235. Single or multiple copies of this article are available for a fee from The Haworth Document Delivery Service [1-800-HAWORTH, 9:00 a.m. - 5:00 p.m. (EST). E-mail address: docdelivery@haworthpress.com].

http://www.haworthpress.com/store/product.asp?sku=J153
© 2003 by Taylor & Francis.
10.1300/J153v05n02_08

INTRODUCTION

Adoption of insect-protected transgenic Bt cotton in Argentina started during the 1998/1999 growing season. About 20,000 ha of land were planted with these varieties in 2000/2001, representing 5% of the total area farmed.

Studies conducted in the US to compare the performance of transgenic Bt varieties vs. conventional ones showed that farmers had more economic benefits when using Bt cotton (Gibson et al., 1997; ReJesus et al., 1997b; Stark, 1997; Weir et al., 1998; Mullins and Mills, 1999; Bryant et al., 1999; Seward et al., 2000; Reed et al., 2000; Karner et al., 2000; Cooke et al., 2000; Adams et al., 2001; Oppenhuizzen et al., 2001).

Studies performed in South Africa show that Bt cotton varieties provide important benefits to small scale farmers, allowing improved insect control, higher yields and considerable economic, health and environmental benefits (Ismael et al., 2001; Ismael et al., 2002; Bennett et al., 2001). Similar observations were made in China, where around 3 million small farmers grew 0.5 million hectares in 2000 (James, 2001; Pray et al., 2001) and in Mexico (Traxler et al., 2001).

Brazil has not approved the commercialization of Bt cotton, but based on field trial results from the 1999/2000 season, simulated scenarios for different situations were developed that indicate an important potential reduction in insecticide use, providing economic, environmental and health benefits for growers (De Souza, Filho, and Gameiro et al., 2002).

Field level data collected from several regions in Argentina from the 1999/2000 and 2000/2001 growing seasons is now available and presented in this report.

Information from 64 commercial plots planted either with Bt cotton or with conventional varieties was compiled during the 1999/2000 growing season. This study was repeated in 2000/2001, in a total of 82 plots. For both growing seasons, the farms were located in Chaco and Santiago del Estero (Figure 1). These two provinces are the main cotton producers of Argentina and constitute about 90% of the area planted with cotton.

To assess the impact of Bt cotton use, plots planted with transgenic varieties were compared to those with conventional varieties, planted by the same farmers in the same areas. Data on inputs and yields, compiled from both growing seasons, are analyzed in this work.

FIGURE 1. Map of Argentina and the 'Argentine cotton belt' region (inset) where the study was performed. Zone 1 includes the sites of *San Bernardo*, *La Clotilde* and *Villa Angela*, while zone 2 encompasses the sites of *Charata*, *Pinedo*, *Campo Largo*, *El Estero*, *Itin* and *Gancedo*; all of them are situated in the Province of Chaco. Zone 3 covers *Frentones* and *Pampa del Infierno* in Chaco, and *Sachayoj* in Santiago del Estero. Zone 4 is *Saenz Peña*, *Napenay*, *Avia Terai* and *Tres Isletas* from Chaco. Zone 5 comprises *Roversi*, *Pozo el Toba*, *Campo del Cielo*, *Quimili*, *Tomas Young* and *Juries*, all situated in Santiago del Estero. Zone 6 includes *El Simbolar*, *La Cañada*, *La Capilla*, *Jumalito*, *Rodeana*, *Fernandez* and *Taboada*, all situated in the area of influence of *La Banda* in Santiago del Estero.

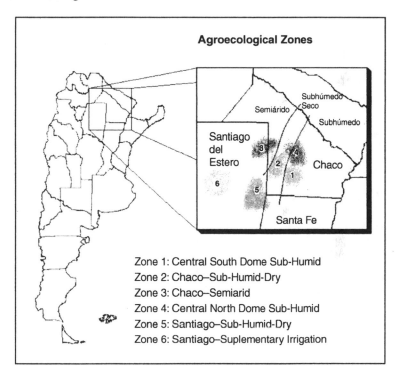

DATA COLLECTION AND STUDY DESIGN

The 1999/2000 growing season was the second in which Bt cotton commercial varieties were commercially planted in Argentina. During this season, a survey was conducted to evaluate the management of these and conventional varieties, in order to detect differences in inputs, costs, yields and income between the two alternatives.

Data collection was carried out in fields planted by 32 farmers in 1999 and fields of 41 farmers in 2000. Each farmer planted one plot with Bt seed and one with conventional seed, side by side. Therefore, 64 and 82 plots were surveyed in 1999 and 2001, respectively, half planted with Bt and half with conventional seed. These plots were distributed along the provinces of Chaco and Santiago del Estero, which constitute the Argentine "cotton belt."

Spreadsheets for field data recording were developed and follow up was done with the help of professionals who provided technical support for insect control in Bt plots and traditional management (by farmers) in conventional ones.

Total acreage for the Bt cotton plots considered in this study was 3,255 hectares (around 33% of the total area planted with Bt cotton in Argentina) in 1999/2000 and 5,256 hectares (or 26% of the total Bt area) in the 2000/2001 season. This study was conducted in different agroecological areas (Figure 1) such as: (a) Central North and South Domes (zones 4 and 1), catalogued according to the Thornthwaite Index (TI) as "sub-humid"; (b) Chaco and Santiago, TI "sub-humid-dry" (zones 2 and 5) and (c) Chaco semiarid (zones 3 and 6), IT "semiarid" (INTA 1990).

While zones 1 and 4 lie on soils originated from *loes* material, the other zones have aluvial soils (Ledesma, 1977).

During the 1999/2000 growing season, the transgenic variety used was Nucotn 33B[1] (NC 33 B) while in 2000/2001, the transgenic varieties used were NC 33 B and DP 50 B,[1] (from now on, all designated "Bt"). Conventional varieties, on the other hand, were Guazuncho 2 INTA,[2] Chaco 520 INTA,[2] Gringo INTA,[2] Pora INTA[2] and DP 5690[1] (from now on, designated "C," or "Conv," or "conventional").

Therefore, all comparisons were done with the above mentioned Bt varieties commercially available in Argentina to date, which are not isogenic (identical genetic background or germplasm) with conventional ones, with the exception of NC33B and DP 5690, which have a similar genetic background. Conventional varieties included in this study are those commonly planted by local growers and the Bt trait is not yet available in these, so some of the observed differences might arise from germplasm or variety differences and may not be attributable to the Bt trait. Therefore, the comparison focused on the adoption vs. non-adoption of the insect protection technology in terms of input expenditures vs. yields. From this standpoint, these results are valid in terms of their effects on final costs and income under real field situations.

The data considered for cost calculations included all labor performed, from initial tillage to the end of harvest. The study considered the frequency and category of labor performed, input costs, and type of harvest. Inputs prices correspond to average values for the Saenz Peña area for both growing seasons.

In order to determine incomes, raw cotton yields per hectare, gin turnout and the commercial grade of the cotton produced were considered. The reference price of cotton used to estimate incomes was the one determined by the Argentine Cotton Chamber for the different qualities and gin turnouts. The final price obtained by farmers is a function of individual gin turnouts and commercial grades. Therefore, each farmer obtained a price based on the final quality and volume of production. In addition, with data from both seasons, we were able to calculate the total amounts of insecticides used in Bt and conventional plots, as well as the total area that was treated with a particular commercial insecticide. This was possible, since the areas and plots planted by each farmer, and the frequency and doses of each application, were recorded. These products were also grouped according the toxicological information of the "Phytosanitary Products Guide for Argentina" (2001 ed., Chamber of Agricultural Health and Fertilizers).

All the reported values, originally expressed in pesos of June for the 1999/2000 growing season and in pesos of April for the 2000/2001 growing season, were adjusted to pesos of February 2002 using the Internal Price Producers Index (PPI) made by the National Institute for Statistics and Surveys (INDEC), base year 1993 = 100, in order to adjust for inflation. Therefore, data compared in this study is expressed in an equivalent currency (February 2002). Until December 2001, the exchange rate was 1 peso = 1 dollar. Nowadays, the dollar has a free ("floating") market value.

METHODS

The method used for this analysis was that of partial budget (Bryant et al., 1997) and marginal approach. According to this method, we considered those items for which it is possible to detect differences in the management of Bt or conventional varieties that translate into different direct costs or incomes. Other components of the fixed costs for managing each plot were not considered, being the same for the different alternatives.

Direct costs determined, considering their separate components, were: tillage labor, crop management, insecticide applications, seed, harvest and marketing. The marginal benefit of each component was estimated as the difference between incomes and marginal costs of Bt vs. conventional varieties (ReJesus et al., 1997a).

Machinery costs, which include the variable costs associated with machinery use, were determined on the basis of a 100 HP tractor with its corresponding tools. These also include removal of the stubble from previous crops, tillage labor, irrigation when needed and agrochemical applications with the exception of insecticides, which were considered separately. In the case of insecticide sprayings, both the labor and the product costs were consolidated.

The cost of each task carried out by the piece of machinery (tractor or other) was determined. These, multiplied by the frequency of use and added up, gave the total machinery cost.

When considering the costs of Bt seed, both the price of the seed and any technology fees made up the final cost. For conventional seed, only the cost of each variety and the amount of seed used by the farmer were considered. Other expenses included herbicides, growth regulators and defoliators, of which the final costs were calculated as the multiplication of the number of applications, doses and prices of the commercial products. The costs of harvest and marketing include: mechanical harvest (contracted); federal, provincial and city taxes; and other expenses generated by the marketing of the product.

The *Total Direct Costs* were obtained adding up all the costs described above. Average values of the results for both varieties were determined considering all the plots in general and also considering each zone separately.

RESULTS

The average results from all fields and the two growing seasons, are shown in Tables 1 to 3, which show individual cost and income components as well as the differences observed for Bt and conventional plots. Detailed data, not shown here, with individual values by zone, are available upon request to the author by e-mail. All figures are in Argentine pesos ($).

The observed differences between Bt and conventional cotton varieties–the number of insecticide applications, gross incomes, total direct costs and differences in net income–for different production zones are

TABLE 1. Cost Components

	1999/2000		2000/2001		AVERAGE	
	Bt n = 32	C n = 32	Bt n = 41	C n = 41	Bt n = 73	C n = 73
Machinery Costs ($/ha)	56.13	55.16	60.12	58.90	57.55	56.46
Insecticide Applications	1.38	3.78	2.38	5.18	1.90	4.49
Insecticide Application ($/ha)	24.83	55.58	16.60	55.29	19.98	54.66
Seed and Technology ($/ha)	105.98	23.50	143.29	42.88	124.98	33.80
Other Expenses ($/ha) (1)	51.04	49.20	63.75	52.05	57.31	50.09
Harvest and Marketing ($/ha)	170.72	120.36	211.45	136.12	190.70	127.35
TOTAL DIRECT COST	**408.70**	**303.80**	**495.21**	**345.25**	**450.51**	**322.35**

TABLE 2. Income Components

	1999/2000		2000/2001		AVERAGE	
	Bt n = 32	C n = 32	Bt n = 41	C n = 41	Bt n = 73	C n = 73
Raw Cotton Yield (kg/ha)	2.082	1.175	2.614	1.607	2.420	1.570
Commercial Grade (more frequent)	C 3/4	C 3/4	C 3/4	D	C 3/4	D
Gin Turnout (%)	32.14	32.68	30.61	32.60	30.86	32.19
Raw Cotton Price ($/t)	263.94	253.42	221.67	223.11	237.17	233.34
GROSS INCOME ($/ha)	**534.46**	**356.94**	**556.96**	**359.28**	**539.47**	**353.34**
GROSS MARGIN ($/ha)	**125.76**	**53.4**	**61.75**	**14.03**	**88.96**	**30.98**

TABLE 3. Difference Between Bt and Conventional Varieties (All Fields)

	1999/2000 n = 32	2000/2001 n = 41	AVERAGE n = 73
Gross Income ($/ha)	177.52	197.68	186.13
Machinery Costs ($/ha)	0.97	1.22	1.09
Insecticide Application Costs ($/ha)	−30.76	−38.69	−34.68
Seed Costs ($/ha)	82.48	100.41	91.18
Other Expenses ($/ha)	1.85	11.70	7.22
Harvest and Marketing Costs ($/ha)	50.36	75.33	63.35
Total Direct Costs ($/ha)	104.90	149.96	128.15
DIFFERENCE IN NET INCOME (Bt-Conv)	**72.62**	**47.72**	**57.98**

compared in Figure 2, A-D. Both combined averages and zonal data are shown. Only zones with two years of data were displayed (individual data for zones 3 and 6 are not shown).

DISCUSSION

The present study was conducted to assess the impact of adoption of Bt cotton varieties in Argentina since their commercial release, in 1998. Data from two years of studies in commercial farmers' fields distributed in the Argentine cotton belt were compared. Only data from zones 1, 2, 4 and 5–from which we have two years (1999/00 and 2000/01) of data–were considered in this report. These data were analyzed collectively and also separately, by zone.

The number of insecticide applications was considerably lower in Bt plots in every zone, and in both seasons (Table 1, Figure 2A). These were between two and three times less (2.59 times in average) than those required for conventional varieties. There was an increase in the number of applications observed between the first and second season for both crop types, probably due to higher insect pressure in general. During the 1999/2000 season, $69,551 was spent to spray all Bt plots, and $100,185 to treat conventional plots. In the last season (2000/2001), $229,275 was spent to treat Bt plots and $389,395 to spray conventional plots, with similar savings for the two seasons (40% and 41%, respectively). Insecticide cost savings for Bt crop use varied between 17.16 and 44.80 $/ha (34.68 $/ha in average).

The applied products were also classified by toxicity and dosage. During the 2000/2001 growing season, milder insecticides were used, classified as of moderate or low toxicity and lower doses were applied per hectare in Bt plots than in conventional ones. In addition, it was observed that more acreage was sprayed with low or mild toxicity products than with highly or extremely toxic products. All the above indicate considerable environmental and health benefits, with lower farmer exposure.

Machinery costs were 2% higher for Bt plots. This is accounted for in the increased growth regulator applications required by DP5415 and NC33B parental varieties (data not shown).

Results showed a trend of increased yields in transgenic varieties (Table 2). Yield increases from 23 to 75% were observed according to zones, which led to higher gross incomes for Bt varieties, between 19%

FIGURE 2. Combined data and data for individual zones (1, 2, 4 and 5) with two seasons of monitoring, covering major input and income features of the study. (A) Average number of insecticide applications. (B) Average gross incomes in $/ha. (C) Average total direct costs for cotton growing, harvesting and processing in $/ha. In A, B and C, Bt cotton (solid bars) is compared to conventional cotton (open bars). (D) Net income difference for Bt versus conventional plots of cotton in zones 1, 2, 4 and 5 (shaded bars) and for the combined analysis (hatched bar), which shows an overall income advantage for Bt cotton use.

A.

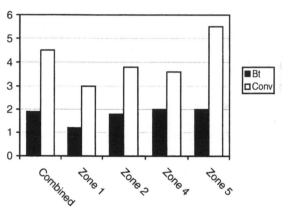

Average Number of Insecticide Applications

B.

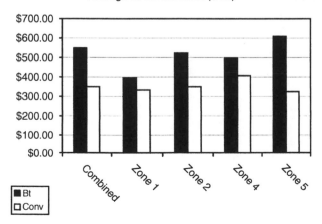

Average Gross Incomes ($/ha)

FIGURE 2 (continued)

C.

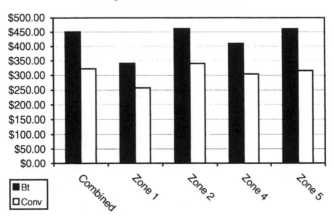

Average Total Direct Costs ($/ha)

D.

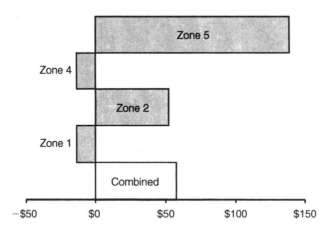

Net Income Difference (Bt − conventional plots) $/ha

(zone 1) and 88% (zone 5). Consequently, gross income increased as well (Tables 2 and 3, Figure 2B).

Differences in seed costs of Bt and conventional varieties may have been a key factor influencing farmer choices. Seed costs increased for both alternatives in the last two years, and conventional seed increased in price by a higher percentage than Bt cotton seed. Bt cotton seed was

4.5 times more expensive than conventional seed in 1999, and 3.3 times more expensive in 2000. The extra cost of Bt cotton seed was higher than the savings coming from the reduced use of insecticide products.

As expected, marketing and harvesting costs were higher for Bt plots (Table 1, Figure 2C) since these costs are directly linked to yields. For this reason, the Total Direct Costs for Bt varieties were 33 to 46% higher than those for conventional crops. All other costs showed differences that did not change the final result.

Although lower insect control costs alone did not compensate for higher seed, harvest and marketing costs for Bt varieties, the impact of higher yields did generate enough income to compensate. Overall, net income was increased (Table 3, Figure 2D), in addition to the environmental and health related benefits derived from applying less chemicals. The net benefit derived from the use of Bt varieties was positive in zone 2 (19%) and zone 5 (88%). In zones 1 and 4, conventional varieties gave a marginally improved net benefit. This is accounted for by either a narrower yield advantage of Bt over conventional plots (zone 1), or with a lower gin turnout of the Bt plots (zone 4), which could be due to the type of varieties that are available with the Bt trait (-1.33% in the combined average and -2.08% for zone 4). In fact, in these zones, farmers preferably planted Guazuncho 2 as the conventional variety, which has shown higher yields and gin turnouts than DP50–the parental variety for DP50 B–in previous trials (O Royo, INTA, personal communication).

Commercial grade (quality) and gin turnout are two of the main components that determine the market price of the product. These differences (mainly attributable to the different germplasms) impacted negatively on the final revenues for Bt vs. conventional varieties. It is also noteworthy that raw cotton prices in the 2000/2001 season were lower than in the previous season.

CONCLUSIONS

After two years of analysis, it is clear that the adoption of Bt cotton varieties in certain areas of Argentina can bring economic benefits for growers as a result of yield increases and insecticide use reduction, even when some items imply higher costs. A shift towards lower toxicity, safer insecticidal products is also observed. The availability of adequate germplasms with the Bt trait seems a key factor to improve the observed benefits.

NOTES

1. Delta Pine or DP varieties.
2. Developed by the National Institute for Agricultural Technology, INTA.

REFERENCES

Adams, L.C., J.J Adamczyk and D.D. Hardee (2001) A two year agronomic evaluation of conventional and transgenic Bt Cotton cultivars in the Mississippi Delta. *Proceedings of the Beltwide Cotton Conference* 2:1031-1033.

Bennett, A.L., A. Bennett, W. Green, C.L.N. Do Toit, E. Richter, L. Van Staden, D. Brits, F. Feiis and J. Van Jaarsveld (2003) Bollworm control with transgenic (Bt) cotton: First results from Africa. *African Entomology*, 2:(in press).

Bryant, K.J., W.C. Robertson and G.M. Lorenz III (1997) Economic evaluation of Bollgard Cotton in Arkansas: *Proceedings of the Beltwide Cotton Conference* 1: 358-359. National Council, Memphis, TN.

Bryant, K.J., W.C. Robertson and G.M. Lorenz III (1999) Economic evaluation of Bollgard® Cotton in Arkansas: *Proceedings of the Beltwide Cotton Conference* 1:349-350.

Cooke, F.T., Jr., W.P. Scott, R.D. Meeks and D.W. Parvin, Jr. (2000) The economics of Bt cotton in the Mississippi Delta–a progress report. *Proceedings Beltwide Cotton Conference* 1:332-334.

De Souza Filho, J.B. and A. Hauber Gameiro (2002) Economics of Bt of Bollgard® cotton in Brazil. *19th Brazilian Entomology Congress.* Manaus/AM-Brazil, *in press.*

Gibson, J.W., D. Laughlin, R.G. Luttrell, D. Parker, J. Reed and A. Harris (1997) Comparison of cost and returns associated with *Heliothis* resistant Bt cotton to non-resistant varieties. *Proceedings of the Beltwide Cotton Conference* 1:244-247. National Council, Memphis, TN.

INTA (1990) Atlas de la República Argentina. *Instituto Nacional de Tecnología Agropecuaria.* Buenos Aires, Argentina.

Ismael, Y., C. Thirtle and L. Beyers (2001) Efficiency effects of Bt cotton adoption by smallholders in Makhathini Flats, KwaZulu-Natal, South Africa. *Proceeding of the 5th International Conference of Biotechnology, Science and Modern Agriculture: A New Industry at the Dawn of the Century,* Ravello, Italy.

Ismael, Y., L. Beyers, L. Lin and C. Thirtle (2002) Smallholder adoption and economic impacts of Bt cotton in the Makhathini Flats, South Africa. *International Consortium on Agricultural Biotechnology Research (ICABR).*

James, C. (2001) Global Review of Commercialized Transgenic Crops. ISAAA Brief No. 24.

Karner, M., A.L. Hutson and J. Goodson (2000) Bollgard-Impact and value to Oklahoma's cotton industry 1996-1999. *Proceedings of the Beltwide Cotton Conference* 2:1289-1293.

Ledesma, L.L. (1977) Introducción al conocimiento de los suelos del Chaco. *Instituto Nacional de Tecnología Agropecuaria (I.N.T.A.).* Estación Experimental Agropecuaria Sáenz Peña, Chaco. Argentina.

Mullins, J.W. and J.M. Mills (1999) Economics of Bollgard versus non Bollgard cotton in 1998. *Proceedings of the Beltwide Cotton Conference* 2:958-961.

Oppenhuizzen, M., J.W. Mullins and J.M. Mills (2001) Six years of economic comparison of Bollgard® cotton. From *Proceeding of the Beltwide Cotton Conference* 2:862-865. National Council, Menphis, TN.

Pray, C.E., J. Huang and F. Qiao (2001) Impact of Bt cotton in China. *World Development* 29(5):1-34.

Reed, J.T., S. Stewart, D. Laughlin, A. Harris, R. Furr and A. Ruscoe (2000) Bt and conventional cotton in the hills and delta of Mississippi: 5 years of comparison. *Proceedings of the Beltwide Cotton Conference* 2:1027-1030.

Rejesus R.M., J.K. Greene, M.D. Hamming and C.E. Curtis (1997a) Economic analysis of insect management strategies for transgenic Bt cotton production in South Carolina. *Proceedings of the Beltwide Cotton Conference* 1:247-251.

Rejesus R.M., R.P.J. Potting, I. Denholm and C.E. Curtis (1997b) Farmers' expectations in the production of transgenic Bt Cotton: Results from a preliminary survey in South Carolina. *Proceedings of the Beltwide Cotton Conference* 1:253-256. National Council, Menphis, TN.

Seward R.W., P.P. Shelby and S.C. Danehower (2000) Performance and insect control cost of Bollgard vs. conventional varieties in Tennessee. *Proceedings of the Beltwide Cotton Conference* 2:1055-1057.

Stark, C.R., Jr. (1997) Economics of transgenic cotton: some indications based on Georgia producers. *Proceedings of the Beltwide Cotton Conference* 1:251-253.

Traxler, G., S. Godoy-Avila, J. Falc-Zepeda and J. Espinoza-Arellano (2001) Transgenic cotton in Mexico. Economic and environmental impacts. <www.Biotech-info-net/Bt_cotton_Mexico.html>.

Weir, A.T., J.W. Mullins and J.M. Mills (1998) Bollgard cotton-update and economic comparison including new varieties. *Proceedings of the Beltwide Cotton Conference* 2:1039-1040.

Index

35S, (see promoter, CaMV 35S)

α-exotoxin, 33
adoption
  Bt crop, 24,32,36,67,88,107,123,
    124,156,157,163,224,226,
    230,233
  IRM, 46,158,168,174
aflatoxin, 125
Agrobacterium tumefaciens, 59,61,63,
  78,81,98,107,179-186,195,
  196,198,204,205
allergic, 7
alternate host, 62,64,145,146,152,
  157-159,171,172,174
Andean (potato) weevil, 95,96,100,
  105,106,108
aphid, 27,61,95
armyworm, 19,41,42,57,121,127,164,
  166,203
  beet (*Spodoptera exigua*; tobacco
    caterpillar), 19,41,63,66,166,
    203
  fall (*S. frugiperda*), 57,121,127,164
  southern (*S. eridania*), 41
  yellow striped (*S. ornithogalli*), 41
Aspergillus flavus, 125

β-exotoxin, 4-6,21
Bacillus anthracis, 4,6
Bacillus cereus, 4
Bacillus sphaericus, 42

Bacillus thuringiensis subspecies
  aizawai, 16,19,34,35T
  israelensis, 16,20,21,33,34,35T,42
  kurstaki, 16,19,25,33,34,35T,43,
    203
  morrisoni (also *tenebrionis*), 16,59
  tenebrionis, 16,33,35T
  tolworthi, 56,58
benefit
  economic, 58,67,106-108,124,224,
    233
  environmental, 13,22,108,123,224,
    230,233
  health, 13,108,224,230,233
beneficial insects/arthropods, 17,20,22,
  23,25,26,32,35,36,43,62,101,
  102,121,123
binding, 4,13,15,16,55,86,126,141,
  162,164-166,210
biocontrol/biological control,
  36,40T,67
biolistic transformation (see also
  microprojectile), 78,81T
biosafety, 54,68,99,106
blackfly (*Simulium damnosum*),
  20,21,33,34,42
Bollgard, 116,126,127,138-141,142T,
  161,163,164,170,171,173
Bollgard II, 46,116,127,165
bollworm
  cotton (*Helicoverpa zea*;
    CBW/CEW), 19,22T,46,56,
    127,139,144,145,147-150,
    150T,151,152T,152-154,
    156-160,162-167,171-173

http://www.haworthpress.com/store/product.asp?sku=J153
© 2003 by Taylor & Francis.
10.1300/J153v05n02_13

Printed and bound by CPI Group (UK) Ltd, Croydon, CR0 4YY

28/10/2024

01780386-0001